# Heterogeneous Catalysts for Petrochemical Synthesis and Oil Refining

# Heterogeneous Catalysts for Petrochemical Synthesis and Oil Refining

Editors

**Eduard Karakhanov**
**Aleksandr Glotov**

MDPI • Basel • Beijing • Wuhan • Barcelona • Belgrade • Manchester • Tokyo • Cluj • Tianjin

*Editors*
Eduard Karakhanov
Petroleum Chemistry and
Organic Catalysis
Lomonosov Moscow State
University
Moscow
Russia

Aleksandr Glotov
Physical and Colloid Chemistry
Gubkin Russian State University
of Oil and Gas
Moscow
Russia

*Editorial Office*
MDPI
St. Alban-Anlage 66
4052 Basel, Switzerland

This is a reprint of articles from the Special Issue published online in the open access journal *Catalysts* (ISSN 2073-4344) (available at: www.mdpi.com/journal/catalysts/special_issues/petrochemical_oil_refining).

For citation purposes, cite each article independently as indicated on the article page online and as indicated below:

LastName, A.A.; LastName, B.B.; LastName, C.C. Article Title. *Journal Name* **Year**, *Volume Number*, Page Range.

**ISBN 978-3-0365-1430-7 (Hbk)**
**ISBN 978-3-0365-1429-1 (PDF)**

© 2021 by the authors. Articles in this book are Open Access and distributed under the Creative Commons Attribution (CC BY) license, which allows users to download, copy and build upon published articles, as long as the author and publisher are properly credited, which ensures maximum dissemination and a wider impact of our publications.

The book as a whole is distributed by MDPI under the terms and conditions of the Creative Commons license CC BY-NC-ND.

# Contents

**About the Editors** . . . . . . . . . . . . . . . . . . . . . . . . . . . . . . . . . . . . . . . . . . . . . . . . . . . . . . . . **vii**

**Aleksandr Glotov and Eduard Karakhanov**
Heterogeneous Catalysts for Petrochemical Synthesis and Oil Refining
Reprinted from: *Catalysts* **2021**, *11*, 602, doi:10.3390/catal11050602 . . . . . . . . . . . . . . . . . . **1**

**Aleksandr Glotov, Anna Vutolkina, Aleksey Pimerzin, Vladimir Nedolivko, Gleb Zasypalov, Valentine Stytsenko, Eduard Karakhanov and Vladimir Vinokurov**
Ruthenium Catalysts Templated on Mesoporous MCM-41 Type Silica and Natural Clay Nanotubes for Hydrogenation of Benzene to Cyclohexane
Reprinted from: *Catalysts* **2020**, *10*, 537, doi:10.3390/catal10050537 . . . . . . . . . . . . . . . . . . **3**

**Leonid Kulikov, Maria Kalinina, Daria Makeeva, Anton Maximov, Yulia Kardasheva, Maria Terenina and Eduard Karakhanov**
Palladium Catalysts Based on Porous Aromatic Frameworks, Modified with Ethanolamino-Groups, for Hydrogenation of Alkynes, Alkenes and Dienes
Reprinted from: *Catalysts* **2020**, *10*, 1106, doi:10.3390/catal10101106 . . . . . . . . . . . . . . . . . . **17**

**Aleksey Pimerzin, Aleksander Savinov, Anna Vutolkina, Anna Makova, Aleksandr Glotov, Vladimir Vinokurov and Andrey Pimerzin**
Transition Metal Sulfides- and Noble Metal-Based Catalysts for N-Hexadecane Hydroisomerization: A Study of Poisons Tolerance
Reprinted from: *Catalysts* **2020**, *10*, 594, doi:10.3390/catal10060594 . . . . . . . . . . . . . . . . . . **35**

**Dmitriy I. Potemkin, Vladimir N. Rogozhnikov, Sergey I. Uskov, Vladislav A. Shilov, Pavel V. Snytnikov and Vladimir A. Sobyanin**
Coupling Pre-Reforming and Partial Oxidation for LPG Conversion to Syngas
Reprinted from: *Catalysts* **2020**, *10*, 1095, doi:10.3390/catal10091095 . . . . . . . . . . . . . . . . . . **51**

**Dmitry Melnikov, Valentine Stytsenko, Elena Saveleva, Mikhail Kotelev, Valentina Lyubimenko, Evgenii Ivanov, Aleksandr Glotov and Vladimir Vinokurov**
Selective Hydrogenation of Acetylene over Pd-Mn/Al$_2$O$_3$ Catalysts
Reprinted from: *Catalysts* **2020**, *10*, 624, doi:10.3390/catal10060624 . . . . . . . . . . . . . . . . . . **59**

**Tatiana Kuchinskaya, Mariia Kniazeva, Vadim Samoilov and Anton Maximov**
In Situ Generated Nanosized Sulfide Ni-W Catalysts Based on Zeolite for the Hydrocracking of the Pyrolysis Fuel Oil into the BTX Fraction
Reprinted from: *Catalysts* **2020**, *10*, 1152, doi:10.3390/catal10101152 . . . . . . . . . . . . . . . . . . **73**

# About the Editors

**Eduard Karakhanov**
Prof. Eduard Karakhanov earned his Ph.D. in Chemistry from Lomonosov Moscow State University in 1963 and received a Doctor of Science (habilitation) in Chemistry in 1977. Since 1983 Prof. Karakhanov has been a chair of the Department of Petroleum Chemistry and Organic Catalysis, Faculty of Chemistry in Lomonosov Moscow State University. His area of specialization is petroleum chemistry, heterogeneous catalysis, oil refining, metal–organic frameworks, dendrimers, zeolites, structured mesoporous aluminosilicates, hydroformylation. Prof. Karakhanov has published more than 450 papers. He has been a scientific advisor of 50 Ph.Ds. Eduard Karakhanov is an Honored Scientist of the Russian Federation. He was awarded the N.D. Zelinsky Prize for outstanding work in the field of organic chemistry and petrochemistry. Prof. Karakhanov is a member of the IUPAC councils of international symposia on macromolecular metal complexes, macro- and supramolecular architecture, and materials.

**Aleksandr Glotov**
Dr. Aleksandr Glotov is a leading researcher, head of the catalysis laboratory in the Gubkin Russian State University of Oil and Gas. He earned his Ph.D. in petroleum chemistry from Lomonosov Moscow State University in 2016. He works on micro/mesoporous functional materials design, including self-assembly, template synthesis and modification of zeolites, ordered mesoporous silicas, aluminosilicate nanotubes for different refining and petrochemical processes (hydroprocessing, isomerization, hydrotreating, aromatics hydrogenation, catalytic cracking).

*Editorial*

# Heterogeneous Catalysts for Petrochemical Synthesis and Oil Refining

**Aleksandr Glotov [1,*] and Eduard Karakhanov [2,*]**

1. Department of Physical and Colloid Chemistry, Faculty of Chemical Technology and Ecology, Gubkin Russian State University of Oil and Gas, 65 Leninsky Prosp., 119991 Moscow, Russia
2. Department of Petroleum Chemistry and Organic Catalysis, Faculty of Chemistry, Lomonosov Moscow State University, 1 Leninskie Gory, 119991 Moscow, Russia
\* Correspondence: glotov.a@gubkin.ru (A.G.); kar@petrol.chem.msu.ru (E.K)

**Keywords:** petrochemical synthesis; oil refining; zeolites; aluminosilicates; organic and metal-organic frameworks; nanotubes; hydroprocessing; C-1 chemistry

In modern industry, more than 90% of processes are catalytic. Heterogeneous catalysis is among the major solutions for cost-effective and sustainable industrial applications and processing. Depending on the refinery and process, the development of heterogeneous catalysts is focused on the increase in feedstock conversion and selectivity to products. The design and development of highly efficient and stable heterogeneous catalysts represent an emerging frontier for overcoming energy and environmental challenges. Many industrial petrochemical and oil refining processes are facing new challenges that can be solved by using heterogeneous catalysts.

Recent trends in heterogeneous catalysis are: the design of new functional composite and nanostructured materials, including traditional zeolites' modification; the synthesis of nanoscale meso/micro materials; the investigation of dispersed nanosized systems; stabilizing and minimizing their aggregation; single-atom catalysis; new strategies for the development of predictable metal-support interaction, induced by a controlled synthesis of active phase and their dispersion. One of the modern routes is metal-organic and covalent frameworks' design with tunable textural and functional properties.

This Special Issue covers the most recent progress and advances in the field of heterogeneous catalysts based on mesoporous composites with embedded halloysite nanotubes covered with ruthenium nanoparticles for exhaustive benzene hydrogenation [1], in situ generated and supported on zeolites' transition metal sulfides for the hydrocracking of the pyrolysis fuel oil and n-alkanes isomerization [2,3]. This issue also includes investigations of novel rhodium systems supported on FeCrAl composite for the coupling of pre-reforming and partial oxidation to liquefied petroleum gas processing into syngas [4]. We have collected works devoted to the palladium catalysts based on porous aromatic frameworks and alumina for the hydrogenation of unsaturated compounds (alkynes, alkenes and dienes) and for the selective removal of acetylene from ethane-ethylene fractions [5,6].

We hope that this Special Issue will be beneficial for researchers working in heterogeneous catalysis and functional materials design, such as zeolite composites, mesoporous materials, transition metal sulfides, aluminosilicates, porous aromatic frameworks, and metal nanoparticles immobilization.

**Author Contributions:** The contributions of A.G. and E.K. are equal. Both authors have read and agreed to the published version of the manuscript.

**Funding:** This research received no external funding.

**Data Availability Statement:** The data that support the findings of this study are available in the literature cited.

**Conflicts of Interest:** The authors declare no conflict of interest.

## References

1. Glotov, A.; Vutolkina, A.; Pimerzin, A.; Nedolivko, V.; Zasypalov, G.; Stytsenko, V.; Karakhanov, E.; Vinokurov, V. Ruthenium Catalysts Templated on Mesoporous MCM-41 Type Silica and Natural Clay Nanotubes for Hydrogenation of Benzene to Cyclohexane. *Catalysts* **2020**, *10*, 537. [CrossRef]
2. Kuchinskaya, T.; Kniazeva, M.; Samoilov, V.; Maximov, A. In Situ Generated Nanosized Sulfide Ni-W Catalysts Based on Zeolite for the Hydrocracking of the Pyrolysis Fuel Oil into the BTX Fraction. *Catalysts* **2020**, *10*, 1152. [CrossRef]
3. Pimerzin, A.; Savinov, A.; Vutolkina, A.; Makova, A.; Glotov, A.; Vinokurov, V.; Pimerzin, A. Transition Metal Sulfides- and Noble Metal-Based Catalysts for N-Hexadecane Hydroisomerization: A Study of Poisons Tolerance. *Catalysts* **2020**, *10*, 594. [CrossRef]
4. Potemkin, D.I.; Rogozhnikov, V.N.; Uskov, S.I.; Shilov, V.A.; Snytnikov, P.V.; Sobyanin, V.A. Coupling Pre-Reforming and Partial Oxidation for LPG Conversion to Syngas. *Catalysts* **2020**, *10*, 1095. [CrossRef]
5. Kulikov, L.; Kalinina, M.; Makeeva, D.; Maximov, A.; Kardasheva, Y.; Terenina, M.; Karakhanov, E. Palladium Catalysts Based on Porous Aromatic Frameworks, Modified with Ethanolamino-Groups, for Hydrogenation of Alkynes, Alkenes and Dienes. *Catalysts* **2020**, *10*, 1106. [CrossRef]
6. Melnikov, D.; Stytsenko, V.; Saveleva, E.; Kotelev, M.; Lyubimenko, V.; Ivanov, E.; Glotov, A.; Vinokurov, V. Selective Hydrogenation of Acetylene over Pd-Mn/$Al_2O_3$ Catalysts. *Catalysts* **2020**, *10*, 624. [CrossRef]

Article

# Ruthenium Catalysts Templated on Mesoporous MCM-41 Type Silica and Natural Clay Nanotubes for Hydrogenation of Benzene to Cyclohexane

Aleksandr Glotov [1,*], Anna Vutolkina [1,2], Aleksey Pimerzin [1,3], Vladimir Nedolivko [1], Gleb Zasypalov [1], Valentine Stytsenko [1], Eduard Karakhanov [2] and Vladimir Vinokurov [1]

1. Gubkin Russian State University of Oil and Gas, 65 Leninsky prosp., 119991 Moscow, Russia; annavutolkina@mail.ru (A.V.); aleksey@pimerzin.com (A.P.); nedolivko74@mail.ru (V.N.); gleb.zasypalov@mail.ru (G.Z.); vds41@mail.ru (V.S.); vinok_ac@mail.ru (V.V.)
2. Lomonosov Moscow State University, 3, 1 Leninskie Gory, 119991 Moscow, Russia; kar@petrol.chem.msu.ru
3. Samara State Technical University, 244 Molodogvardeyskaya street, 443100 Samara, Russia
* Correspondence: glotov.a@gubkin.ru

Received: 13 April 2020; Accepted: 11 May 2020; Published: 13 May 2020

**Abstract:** Mesoporous ruthenium catalysts (0.74–3.06 wt%) based on ordered Mobil Composition of Matter No. 41 (MCM-41) silica arrays on aluminosilicate halloysite nanotubes (HNTs), as well as HNT-based counterparts, were synthesized and tested in benzene hydrogenation. The structure of HNT core-shell silica composite-supported Ru catalysts were investigated by transmission electron microscopy (TEM), X-ray fluorescence (XRF) and temperature-programmed reduction (TPR-H$_2$). The textural characteristics were specified by low-temperature nitrogen adsorption/desorption. The catalytic evaluation of Ru nanoparticles supported on both the pristine HNTs and MCM-41/HNT composite in benzene hydrogenation was carried out in a Parr multiple reactor system with batch stirred reactors (autoclaves) at 80 °C, a hydrogen pressure of 3.0 MPa and a hydrogen/benzene molar ratio of 3.3. Due to its hierarchical structure and high specific surface area, the MCM-41/HNT composite provided the uniform distribution and stabilization of Ru nanoparticles (NPs) resulted in the higher specific activity and stability as compared with the HNT-based counterpart. The highest specific activity (5594 h$^{-1}$) along with deep benzene hydrogenation to cyclohexane was achieved for the Ru/MCM-41/HNT catalyst with a low metal content.

**Keywords:** ruthenium catalysts; benzene hydrogenation; MCM-41; halloysite nanotubes; mesoporous aluminosilicates; MCM-41/HNT composite

## 1. Introduction

In the modern global quest for cleaner fuel production, benzene has been identified as a gasoline component that should be reduced. According to the modern clean fuel standard regulations in the US, specifically Mobil Source Air Toxics II (MSAT II), refiners are required to reduce benzene in gasoline to 0.62 vol% on an average annual basis. In Europe and in many other regions, a regulation of 1.0 vol% maximum of benzene in gasoline has also been adopted to limit benzene [1–3]. The selective removal of benzene and other aromatics from motor fuels by hydrogenation ensures a control of the particulate emissions and cetane number boost of diesel [2].

There are two main strategies for benzene hydrogenation: partial hydrogenation aimed to cyclohexene production and deep hydrogenation to cyclohexane [4–7]. The former often requires bimetallic systems as catalysts, such as Ru–Zn, Ru–Co, Ru–Cu and Ru–lanthanides, promoted by various additives or non-promoted [4,7–9]. Another approach is to design hydrophobic/hydrophilic supports and use water as a solvent to prevent cyclohexene excess by hydrogenation (water solubility of cyclohexene and benzene is higher than that of cyclohexane alone) [4,10].

In refineries, benzene hydrogenation to cyclohexane performs in the presence of Group VIII metal catalysts such as Ni, Pt, Pd and Ru at temperatures in the range 150 °C–220 °C under $H_2$ pressure up to 10 MPa [11–16]. For deep benzene hydrogenation at lower temperatures, Ni- or Pt-containing catalysts are usually employed as they are the most active [11,17–19]. These catalysts, however, have a low tolerance to various poisons in the feed that should be preliminary refined [18–20]. The conventional sulfide NiMoSx and NiWSx catalysts have a good activity only at severe conditions (T > 300 °C and pressures 5 MPa and higher), therefore, sophisticated equipment is needed, causing higher investment and process costs [19,21,22].

The needs for effective catalysts ensuring benzene removal under mild conditions have initiated a number of studies aimed at designing new catalytic systems comprising noble metals on supports such as alumina, zeolites and ordered mesoporous silica, e.g., SBA-15, Mobil Composition of Matter No. 41 (MCM-41) [11,17–19,23–25]. There are reports on the application of Ru/C, Ru/graphene and Ru/carbon nanotubes in deep benzene hydrogenation [9,26]. Some research used bimetallic systems with noble metals [23,27,28].

The most promising catalytic systems for benzene hydrogenation to cyclohexane are those comprising ruthenium, thanks to its high activity and low cost compared to other noble metals [25,29–31]. The prospects stated have been confirmed by the application of Ru/zeolite catalysts for benzene hydrogenation [25,32]. In the operation of zeolite-based catalysts, however, diffusion limitations arise due to very narrow channels inside this support. Thanks to the intrinsic well ordered structure, a high specific surface area (about 1000 $m^2 \cdot g^{-1}$) and adjusting pore sizes (2–4 nm), mesoporous silicates are considered as the promising supports for highly dispersed catalysts [33–36]. Among them, MCM-41 (Mobil Composition of Matter No. 41) is an advanced mesoporous material, with a hierarchical hexagonal 2D structure belonging to the silicate family, and is the most attractive [37–39]. The mean size of MCM-41-supported Ru particles is about 1.8 nm with a metal dispersion of 62% [40]. However, MCM-41 silica possesses a low thermal stability (700°C) and mechanical strength (about 220 MPa) that restricts its industrial application [34,41].

As the carriers for hydrogenation catalysts, natural clay nanotubes such as halloysite are of particular interest [16,42–45]. Halloysite is a natural clay aluminosilicate nanotube from the kaolinite group named after the Belgian geologist Omalius d'Halloy who was the first to describe the mineral. Halloysite nanotubes (HNTs) form by the rolling of the kaolin sheets into tubes (length of 0.5–2 µm, inner diameter 10–30 nm, depending on deposit) with negative an outer surface (tetrahedral silanol groups) and an octahedral alumina-composed positive charged inner surface [42,46]. Halloysite has the appropriate specific surface area (50–100 $m^2 \cdot g^{-1}$), high ion-exchange capacity and mesoporous structure that enables the synthesis of highly active ruthenium catalysts and new materials applied for heterogeneous catalytic systems [15,47,48]. Thus, a new approach was developed, where HNTs act as a template for the self assembling of the mesoporous silica MCM-41 type on the outer surface of HNTs. As a result, high-porous meso silica arrays on HNTs with enhanced thermal and mechanical stabilities were formed [49].

The present work was devoted to the catalytic evaluation of ruthenium catalysts based on ordered MCM-41 type silica arrays on aluminosilicate HNTs in comparison with a HNT-based counterpart, depending on the Ru content as well as the localization of active metal particles in benzene hydrogenation.

## 2. Results and Discussion

The structure of the well ordered mesoporous MCM-41 type silica assembled on the outer surface of HNTs, retained after Ru loading, was clearly indicated by transmission electron microscopy (TEM) (Figure 1). The core-shell hexagonal 2D structure produced by the cetyltrimethylammonium bromide (CTAB)-templated silica on the nanotubes as well as its Ru deposition are shown in Figure 1.

The mesoporous silica phase with a one-direction channel system bonded to the outer surface of aluminosilicate HNTs was kept during the metal loading under microwave irradiation [49].

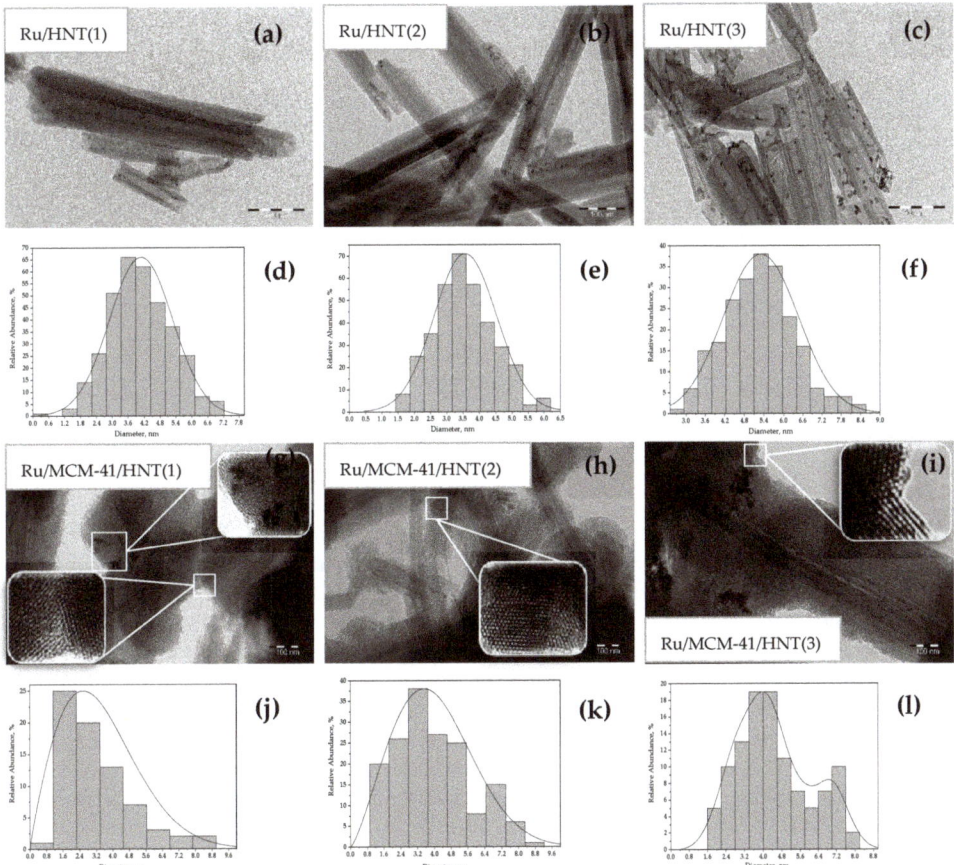

**Figure 1.** Transmission electron microscopy (TEM) micrographs (**a**–**c**,**g**–**i**) and the Ru-nanoparticle size distribution (**d**–**f**,**j**–**l**) in catalysts Ru/ halloysite nanotubes (HNT) and Ru/Mobil Composition of Matter No. 41 (MCM-41)/HNT.

The retention of the MCM-41/HNT structure after Ru impregnation was also proved by low-temperature nitrogen adsorption/desorption technique. As shown in Figure 2, Ru/MCM-41/HNT catalysts were characterized by isotherms of IV type with a capillary condensation step in the range of relative partial pressures $P/P_0$ of 0.4–0.6, corresponding to the presence of a mesoporous framework (Figure 2a) [49,50]. Meanwhile, the $N_2$ isotherms for the Ru/HNT catalysts are of III type with a hysteresis loop at a $P/P_0$ ratio of 0.5–1.0, indicating a capillary condensation in the meso/macropores of halloysite lumen (Figure 2b) [15,16,29,45]. The pore size distribution for both the mesoporous MCM-41/HNT composite-supported and the pristine HNT-based Ru catalysts had a narrow peak centered at 30–32 Å (Figure 2, Table 1).

**Table 1.** Ru content (from X-ray fluorescence (XRF) and temperature-programmed reduction (TPR-H$_2$)) and the textural characteristics of MCM-41/HNT and HNT-supported catalysts.

| Sample | Textural Characteristics | | | Ru, wt% | | | Ru Average Particle Size, nm ** |
|---|---|---|---|---|---|---|---|
| | $S_{BET}$, m$^2 \cdot$g$^{-1}$ | $D_p$, Å | $V_p$, cm$^3$ g$^{-1}$ | From XRF | From TPR-H$_2$ | From XRF (Recycled) * | |
| HNT | 70 | 70 | 0.16 | - | - | - | - |
| Ru/HNT(1) | 68 | 69 | 0.14 | 0.74 | 0.80 | 0.62/0.62/0.61 | 3.6 ± 0.1 |
| Ru/HNT(2) | 61 | 68 | 0.13 | 1.65 | 1.68 | 1.29/1.28/1.27 | 4.1 ± 0.1 |
| Ru/HNT(3) | 56 | 67 | 0.11 | 2.82 | 2.75 | 2.14/2.12/2.09 | 5.4 ± 0.1 |
| MCM-41/HNT | 520 | 28 | 0.43 | - | - | - | - |
| Ru/MCM-41/HNT(1) | 433 | 30 | 0.34 | 0.74 | 0.82 | 0.68/0.68/0.67 | 3.4 ± 0.1 |
| Ru/MCM-41/HNT(2) | 411 | 31 | 0.33 | 1.59 | 1.50 | 1.49 | 3.7 ± 0.1<br>6.7 ± 0.1 |
| Ru/MCM-41/HNT(3) | 373 | 33 | 0.31 | 3.06 | 2.96 | 2.83 | 3.4 ± 0.1<br>7.2 ± 0.2 |

* Measured by XRF after 1 recycle for Ru/MCM-41/HNT (2 and 3) and after each of 3 recycles for other catalysts, ** According to TEM data.

The relative intensity of the shoulder in the range of 20–30 Å (Figure 2b,d) increased with the metal loading, which indicated the partial agglomeration of the Ru nanoparticles inside the pores leading to a decrease in pore volume. It also corresponded with the Ru average particle size calculated based on the TEM data. Thus, for both the MCM-41/HNTs and the pristine HNT-templated catalysts, the pores with more than 40 Å in diameter corresponding to HNTs were also observed [15,30]. The higher metal content and the lower area was found under the part of the pore size distribution curve between 40–80 Å due to the Ru loading into the lumen that was also depicted in the TEM images.

When the metal content increased, the specific surface area of the catalysts decreased. Thanks to the well ordered MCM-41 hexagonal porous arrangement, the $S_{BET}$ for the composite-supported Ru catalysts was significantly higher compared to the pristine HNT-based counterparts. The impregnation of supports with an aqueous solution of ruthenium salt under microwave irradiation provided highly dispersed catalysts having metal nanoparticles being uniformly distributed over the carrier surfaces and in the lumen [15,45] (Figure 1). The impregnation procedure applied gave rise to the forming ruthenium nanoparticles with diameters of 3.6–5.4 nm in the inner surface of the halloysite. This fact was unusual because the positive charge of the lumen normally prevents ruthenium cation intercalation (Figure 1) [29,42,46]. Increasing the metal content leads to then formation and aggregation of partially outside nanoparticles, in accordance with the data published [16,47,48,51]. The average particle size for the Ru/HNT(1) catalyst was about 3.6±0.1 nm and the size distribution curves were approximated by Weibull distribution. For two other counterparts with 1.7 and 2.8wt% of Ru the particles with 4.1 ± 0.1 and 5.4 ± 0.1 nm in diameter were formed, respectively (Table 1, Figure 1). Thus, the higher the metal loading, the higher are the sizes of the particles observed. Moreover, the particle size distributions curves were broadened. The well ordered MCM-41 hexagonal porous arrangement provided uniform particle size distribution. The unimodal distribution with an asymmetric peak for catalysts with a lower Ru content proved the selective nanoparticle intercalation into the mesoporous structure (3.4 ± 0.1 nm), while for Ru/MCM-41/HNT(2) and Ru/MCM-41/HNT(3), bimodal distribution was realized with particle diameters of 3.4 and 6.7–7.2 ± 0.1 nm, respectively, and Ru was intercalated into lumen.

The TPR-H$_2$ profiles for the Ru/HNT samples and the quantification data are presented in Figure 3 and Table 1, respectively. The ruthenium content was calculated based on the hydrogen consumption caused by the complete reduction of RuO$_2$.

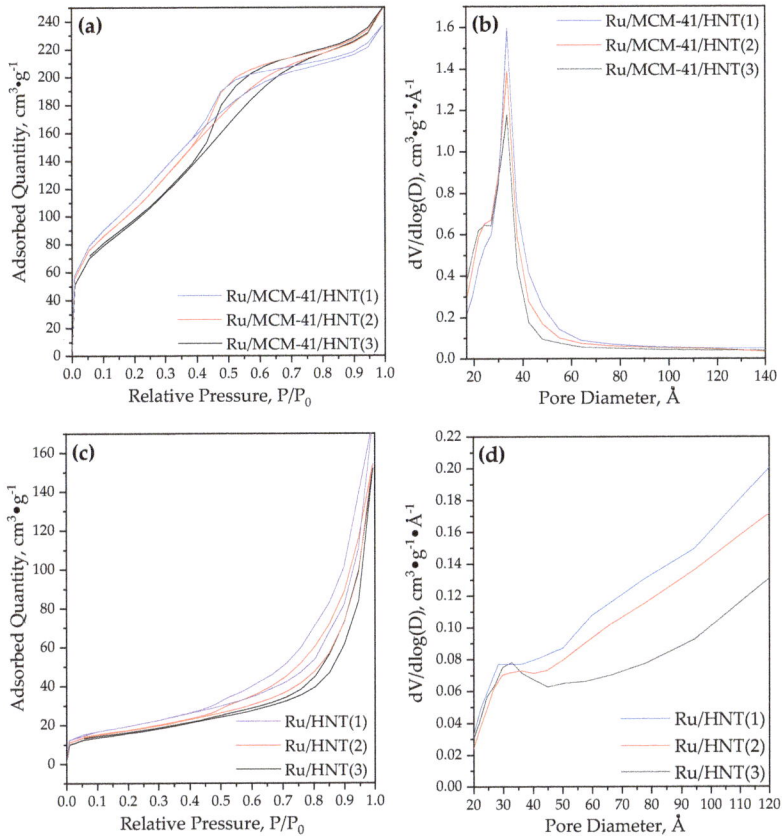

**Figure 2.** Nitrogen adsorption/desorption isotherms (**a**,**c**) and the pore size distribution (**b**,**d**) for the MCM-41/HNT and the HNT-supported Ru catalysts.

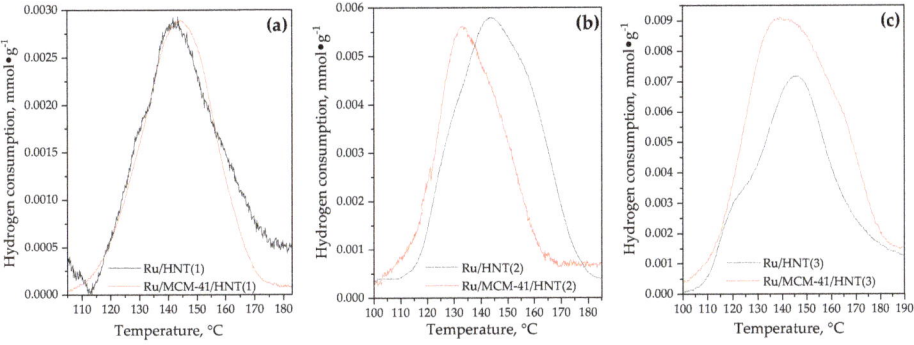

**Figure 3.** TPR-H$_2$ profiles for the Ru catalysts templated on HNTs and the MCM-41/HNT composite (**a**) Ru/HNT(1) and Ru/MCM-41/HNT(1), (**b**) Ru/HNT(2) and Ru/MCM-41/HNT(2), (**c**) Ru/HNT(3) and Ru/MCM-41/HNT(3).

The TPR profiles of the Ru/HNT catalyst had one intense peak at 140 °C corresponding to the reduction of RuO$_2$, which was strongly bonded to the outer surface of the nanotubes [15,30]. An increase in the metal content followed by a partial Ru intercalation into the HNTs lumen, broadens the peak and shifts it to a higher temperature. Thus, the shoulder at a temperature in the range of 155 °C–157 °C was ascribed to the reduction of $RuO_2$ particles formed inside the nanotubes [30]. For the Ru/HNT(2) catalyst, the peak broadening at about 120 °C as well as a small shoulder in the TPR-$H_2$ profile for the Ru/HNT(3) may correspond to the reduction of either agglomerated or physically adsorbed $RuO_2$ nanoparticles. For the MCM-41/HNT-based catalyst, with a low Ru content the peak in the TPR-$H_2$ profile was the same as for the HNT-templated counterpart. Thus, it should be concluded that the Ru nanoparticles were located preliminary outside the mesopores. When the Ru content increased, peaks in the TPR-$H_2$ profile held stable and the peaks were symmetrical but broadened. The shoulder at 130 °C was ascribed to the Ru nanoparticles physically adsorbed outside of both the MCM-41 and the HNTs pores [52]. Meanwhile, for the Ru/MCM-41/HNT(3) sample it may have been caused by particle agglomeration. On the right side from the mean center of the curve, at 165 °C the reduction of the nanoparticles incorporated into the porous MCM-41 structure occurs, which strongly bonded to the support's surface.

The catalytic properties of the samples obtained were compared for the hydrogenation of benzene. They were evaluated as the specific activity ($A_{sp}$) calculated from initial activity (mol (benzene)/mol Ru per hour) and the final benzene conversion at the end of the test (180 min). The results for the hydrogenation of benzene are summarized in Figure 4 and Table 2. Cyclohexane was the only product of benzene hydrogenation over all the Ru/HNT and Ru/MCM-41/HNT catalysts.

As depicted in Figure 4, the final benzene conversion over the catalysts based on MCM-41/HNT was higher in all the tests excluding the samples with 3%wt. Ru content. Meanwhile, the Ru/HNT(2) and the Ru/HNT(3) samples had comparable specific activities with those obtained in the composite-based counterparts (Table 2). It may be due to the partial Ru agglomeration on the external surface of HNTs being more available for benzene molecules, while for MCM-41/HNT-supported catalysts the Ru nanoparticles were incorporated into a well ordered silica porous system, which needed time for the diffusion of reagents to the active sites (Figure 1). For the Ru/HNT(2) and the Ru/MCM-41/HNT(2) composites the specific activities were 1856 and 2079 $h^{-1}$, respectively, while the final benzene conversion over the halloysite-based catalysts was lower compared to the composite-supported ones. In the case of the Ru/MCM-41/HNT(3) and the Ru/HNT(3) catalysts, the ruthenium nanoparticles were partially deposited in the lumen of the halloysite leading to their comparable activity (1535 vs. 1492 $h^{-1}$). This difference in activities may have been caused by the lower dispersion of the active phase over the surface of the MCM-41/HNT composite, having the high content of large Ru nanoparticles. Ru/MCM-41/HNT(1), with an average particle size of 3.4 ± 0.1 nm being uniformly distributed over the surface of mesoporous aluminosilicate support, was the most active with $A_{sp}$ = 5594 $h^{-1}$ (Figure 4a, Table 2). Thus, benzene conversion over this catalyst exceeded 90% in 45 min and attained 100% in 90 min. This tendency was maintained for the calculated specific normalized activity (Table 2).

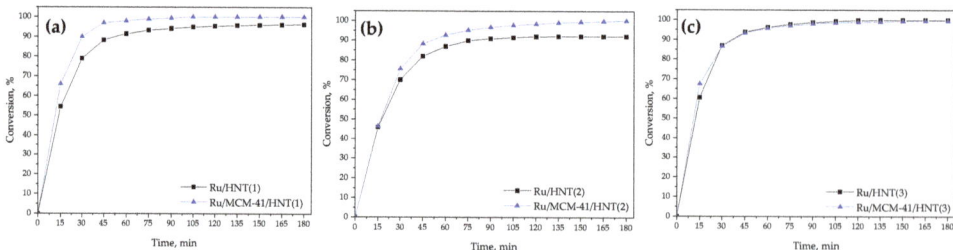

**Figure 4.** Benzene hydrogenation over the HNTs and the MCM-41/HNT-supported Ru catalysts (temperature 80 °C, hydrogen pressure 3MPa, $H_2$/substrate molar ratio of 3.3). (**a**) Ru/HNT(1) and Ru/MCM-41/HNT(1), (**b**) Ru/HNT(2) and Ru/MCM-41/HNT(2), (**c**) Ru/HNT(3) and Ru/MCM-41/HNT(3).

**Table 2.** Catalytic properties of the fresh and recycled Ru-catalysts based on the HNT and the MCM-41/HNT composite.

| Sample | | $A_{sp}$, $h^{-1}$ | $A_{spn}$ *, mol·m$^{-2}$·h$^{-1}$ | Final Benzene Conversion, % |
|---|---|---|---|---|
| Fresh | Ru/HNT(1) | 4610 | 0.34 | 92 |
| | Ru/HNT(2) | 1856 | 0.16 | 96 |
| | Ru/HNT(3) | 1492 | 0.17 | 100 |
| | Ru/MCM-41/HNT(1) | 5594 | 0.39 | 100 |
| | Ru/MCM-41/HNT(2) | 2079 | 0.22 | 100 |
| | Ru/MCM-41/HNT(3) | 1535 | 0.17 | 100 |
| After 3rd recycle | Ru/HNT(1) | 1986 | n.a. | 71 |
| | Ru/HNT(2) | 1680 | n.a. | 74 |
| | Ru/HNT(3) | 580 | n.a. | 75 |
| | Ru/MCM-41/HNT(1) | 4667 | n.a. | 95 |
| After 1st recycle | Ru/MCM-41/HNT(2) | 2064 | n.a. | 98 |
| | Ru/MCM-41/HNT(3) | 1384 | n.a. | 99 |

* calculated from Ru average particles size for the bimodal distribution.

As mentioned above, the most active catalysts were Ru/MCM-41/HNT(1) and Ru/HNT(1) with $A_{spn}$ 0.39 and 0.34 mol·h$^{-1}$·m$^{-2}$, respectively.

It should be noted that for ruthenium catalysts based on MCM-41/HNT composite (samples 2 and 3), the decrease in the Asp and Aspn (Asp normalized per ruthenium specific surface area) parameters was close (24% and 26%, respectively), which was caused by the similar average NPs diameters, while the Ru content became the major factor influenced by catalytic activity (Tables 1 and 2). In the case of Ru/HNT(2), it was calculated that $A_{sp}$ as well as $A_{spn}$ decreased more than twice as compared to the Ru/HNT(1) counterpart. As for Ru/HNT(2) and Ru/HNT(3), the $A_{spn}$ parameters were close, while Asp was reduced by 20%, confirming the negative enlargement effect of nanoparticles on a decrease in catalytic activity [53].

For the supported metal catalysts, stability, as well as selectivity and activity, are the key factors for its further industrial application. From this point of view, we performed stability tests under the same conditions as compared to the fresh catalysts. As depicted in Figure 5a–c, after recycling, the activity of HNT-based catalysts decreased. It should be noted that the greatest reduction in activity occurred after the first cycle for all the catalysts probably caused by the ruthenium leaching. After the first cycle, the ruthenium content decreased by more than 10% from its initial value and remained practically unchanged for the second and third cycles (Table 1). It was also proved by similar kinetic curves for all the Ru/HNT catalysts, in addition to the final benzene conversions almost being the same (70%–80 %) (Figure 5a–c, Table 2).

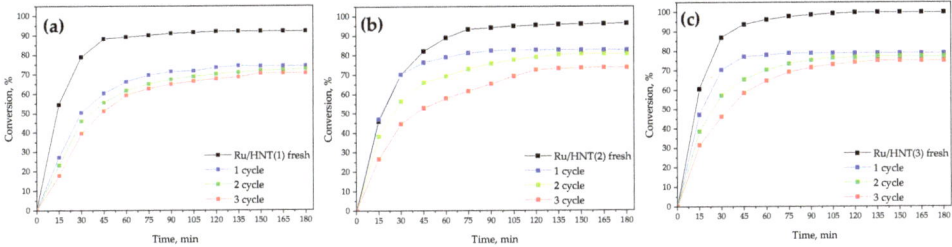

**Figure 5.** Recycle tests for the HNT-supported Ru catalysts (temperature 80 °C, hydrogen pressure 3MPa, H$_2$/substrate molar ratio of 3.3). (**a**) Ru/HNT(1), (**b**) Ru/HNT(2), (**c**) Ru/HNT(3).

Another finding was observed for the catalysts supported on the MCM-41/HNT composite. For all the samples, partial leaching with lower rates compared to the HNT-supported systems was found (Table 1). This may be due to the hierarchical structure and high specific surface area of MCM-41/composite-based catalysts, which resulted in the stabilization of ruthenium nanoparticles within the porous system. As a result, the specific catalytic activity and the final benzene conversion for Ru/MCM-41/HNT(1) after three cycles were comparable with those obtained on the fresh Ru/HNT(1) (Table 2, Figures 5 and 6).

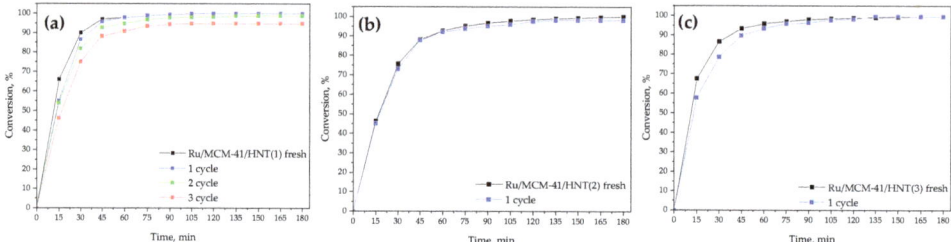

**Figure 6.** Recycle tests for the MCM-41/HNT-composite supported Ru catalysts (temperature 80 °C, hydrogen pressure 3MPa, $H_2$/substrate molar ratio of 3.3). (**a**) Ru/MCM-41/HNT(1), (**b**) Ru/MCM-41/HNT(2), (**c**) Ru/MCM-41/HNT(3).

Finally, we compared the best Ru/HNT(1) and Ru/MCM-41/HNT(1) catalysts with the other systems represented in the literature for benzene hydrogenation (Table 3). As can be seen from Table 3, the specific catalytic activities of the investigated catalysts were much higher compared to the other ruthenium-containing catalysts, based on their supports of a different nature, reported in the literature.

**Table 3.** Comparison of the activity for Ru/HNT(1) and Ru/MCM-41/HNT(1) with the different catalysts reported in the literature.

| Sample | TOF, $h^{-1}$ | Ru, wt% | Benzene/Ru Molar Ratio | Benzene Conversion, % | Time, Min | Sel. to Cyclohexane | Temperature, °C; | $P(H_2)$, MPa | Reference |
|---|---|---|---|---|---|---|---|---|---|
| Ru/C | 1600 | 4.9 | 2000 | 100 | 75 | 100 | 110 | 4 | [54] |
| Ru/Al$_2$O$_3$ | 1416 | 4.0 | 1400 | 100 | 60 | 100 | 80 | 2 | [55] |
| Ru/CNTs | 649 | 4.0 | 500 | 53 | 60 | 98 | 70 | 1 | [56] |
| Ru/montmorillonite | 270 | 0.83 | 275 | 100 | 60 | 100 | 100 | 3,5 | [57] |
| Ru/PAFs | 1600 | 4.83 | 2000 | 79 | 60 | 100 | 80 | 3.3 | [58] |
| Ru/MOFs | 3478 | 5.0 | 8000 | 100 | 135 | 100 | 160 | 6 | [59] |
| Ru/TEGO* | 1302 | 4 | 500 | 75 | 30 | 100 | 70 | 1 | [9] |
| Ru/hydrotalcite | 1300 | 1 | 1300 | 100 | 60 | 100 | 120 | 6 | [60] |
| Ru/HNT(1) | 4610 | 0.74 | 2300 | 92 | 90 | 100 | 80 | 3 | this work |
| Ru/MCM-41/HNT(1) | 5594 | 0.74 | | 100 | 75 | | | | |

* TEGO-thermally exfoliated graphite oxide.

It can be concluded that catalysts based on mesoporous MCM-41/HNT have both the higher hydrogenation activity and stability in benzene hydrogenation to cyclohexane. Most probably, it is caused by the extremely high specific surface area of the MCM-41/HNT support as compared with the HNT (as high as 5–8 times, see Table 1). It should be also noted that for all samples based on HNT, as well as on MCM-41/HNT, the higher is metal loading is, the more the catalytic activity decreases. The higher hydrogenation activity of the catalysts with Ru-loading < 1 wt% is explained by the forming of highly dispersed metal particles on the carrier surface. In the case of Ru-loading in the range of 1.3–3 wt%, the metal particles are larger within the partial agglomeration and the distribution thereof is non-uniform (Figure 1). When the metal content is high, the Ru crust forming egg-shell may occur resulting in the blockage of active sites, and the catalysts' specific activity decreasing.

## 3. Materials and Methods

### 3.1. Chemicals

The following chemicals were used for the synthesis of catalysts and as reference compounds for the gas chromatography analysis: ruthenium (III) chloride (high purity grade, Aurat, Moscow, Russia), halloysite nanoclay (≥98%, Sigma-Aldrich, St. Louis, MO, USA), hexadecyltrimethylammonium bromide (≥98%, Sigma-Aldrich, St. Louis, MO, USA), tetraethyl orthosilicate (≥98%, Sigma-Aldrich, St. Louis, MO, USA), benzene (≥99%, chemical grade, ECOS-1, Moscow, Russia), cyclohexane (for gas chromatography, Supelco, St. Louis, MO, USA), boric acid (purum, ChemMed, Moscow, Russia).

Double distilled water, ethanol and isopropanol (Reachim, Purum, Moscow, Russia) were used as solvents, and ammonia hydroxide (~25%, ECOS-1, Moscow, Russia) was used for adjusting the pH in the MCM-41/halloysite synthesis.

### 3.2. The Synthesis of Catalysts

The ordered mesoporous composite MCM-41/HNT was prepared by the template synthesis, using CTAB as a structure-directing agent for the MCM-41 phase formation [49,61]. The ruthenium deposition was performed by the incipient wetness impregnation technique under microwave irradiation as follows. MCM-41/HNT (1 g) powder was dispersed in a water solution (40 mL) of $RuCl_3$ in the required amount to obtain appropriate metal loading. The dispersion obtained was placed in an ultrasonic bath for 30 min followed by microwave irradiation (800 W) for 3 min and centrifugation (7000 rpm for 2 min). The precipitate was separated and treated with an aqueous solution (30 mL) of $NaBH_4$ (0.5 M) for the ruthenium reduction. The resulting materials were washed with distilled water, separated via centrifugation and dried at 60 °C for 24 h. The finished solid catalysts were grinded into powder and denoted as Ru/MCM-41/HNT(1), Ru/MCM-41/HNT(2) and Ru/MCM-41/HNT(3) according to their ruthenium content. The ruthenium deposition over the HNTs was performed by the intrinsic wetness procedure under microwave irradiation described in [8,19,20,35].

### 3.3. Analyses and Instrumentations

The ruthenium content was determined using an ARL Perform'X X-ray fluorescence spectrometer (Thermo Fisher Scientific, Waltham, MA, USA). The analysis was performed in vacuum using the UniQuant (Thermo Fisher Scientific, Waltham, MA, USA) program without a standard. Before the analysis, the sample was pressured in tablets with boric acid.

Transmission electron microscopy (TEM) images were obtained on a Jem-2100 (JEOL, Tokyo, Japan) microscope with an accelerating voltage of 100 kV. The sample analyzed was ultrasonically dispersed in ethanol. The particle size distribution was obtained by a statistical evaluation of around 1000 particles from different areas of a number of various TEM images using Image-Pro Plus 6.0 software.

The textural properties of the synthesized and catalyst materials, such as the specific surface area ($S_{BET}$), volume ($V_p$) and diameter ($D_p$) of the pores were determined by a low temperature nitrogen adsorption/desorption technique using a Gemini VII 2390t (Micromeritics Instrument Corp., Norcross, GA, USA) instrument at a temperature of 77 K. Before measurements, the samples were outgassed in vacuum at 300 °C for 4 h. The specific surface was calculated according to the Brunauer–Emmett–Teller (BET) equation in a relative pressure range from 0.04 to 0.25 of the adsorption data. The volume of the pores and their diameter were estimated in terms of the Barrett–Joyner–Halenda model (data obtained from the desorption branch of the isotherm).

Temperature-programmed reduction with hydrogen (TPR-$H_2$) was performed with a AutoChem 2950HP instrument (Micromeritics Instrument Corp., Norcross, GA, USA). Before the analysis, the catalyst (100 mg) was pretreated at 400 °C for 30 min under air flow to oxidize the ruthenium nanoparticles. Then, a sample was purged with Ar flow at 400 °C for 1 h, cooled to 50 °C. The reduction step was performed under the 30 mL/min flow of 8 vol.% $H_2$–92 vol.% Ar mixture in the range of 50 °C–400 °C with a ramp of 10 K/min. The consumption of H$_2$ and the ruthenium content were calculated using the AutoChem HP V2.04 program (Micromeritics Instrument Corp., Norcross, GA, USA).

### 3.4. Catalytic Experiments

The catalytic activity in benzene hydrogenation was evaluated in a Parr 5000 Multiple Reactor System (Parr Instruments, Frankfurt am Main, Germany) with stainless steel batch reactors having a Teflon inlet and a magnetic stirrer. The reactor was loaded with benzene (0.01 moles) without any solvent and 60 mg of the catalyst, purged with hydrogen and then pressurized. The catalytic tests were run with 1500 rpm stirring at a hydrogen pressure of 3.0 MPa and a temperature of 80 °C. After reaction, the reactor was cooled down to room temperature, the pressure was dropped to atmospheric and the catalyst was removed from the reaction products via centrifugation, washed 3 times with 10 mL of ethanol and dried at 60 °C for 24 h. The recycling tests were performed under the same conditions as for fresh catalysts. The hydrogenation products were analyzed in isotherm (110 °C) using a Chromos GC-1000 gas chromatograph (Chromos Engineering, Dzerjinsk, Russia) equipped with a flame-ionization detector and a capillary column MEGA-WAX Spirit (MEGA, Legnano, Italy).

The specific catalyst's activity in hydrogenation ($A_{sp}$, h$^{-1}$) was calculated as the amount of reacted benzene (Nb*Cb) per mole of ruthenium and the time in hours, according to the formula:

$$A_{sp} = \frac{N \times Cb}{m_{cat} \times \frac{\omega_{Ru}}{M_{Ru}} \times t_i},$$

where Nb—moles of benzene, Cb—benzene conversion, $m_{cat}$—the catalyst weight, $\omega_{Ru}$—ruthenium content determined by XRF, $M_{Ru}$—ruthenium molar mass and $t_i$ is the time for which the benzene conversion (Cb) was evaluated.

The specific catalyst's activity normalized per ruthenium specific surface area ($A_{spn}$) was calculated as follows:

$$A_{spn} = \frac{A_{sp}}{S_{sp} \times M_{Ru}},$$

where $A_{sp}$—specific catalyst's activity and $S_{sp}$—ruthenium specific surface area, calculated as:

$$S_{sp} = \frac{6}{d \times \rho_{Ru}},$$

where d—average nanoparticle diameter, ρ—metal density, and k—shape factor (6 for spherical nanoparticles).

Each experiment was carried out three times under the same conditions, with the results differing by no more than 2% from the corresponding average value. The measurement error did not exceed 1%.

## 4. Conclusions

Mesoporous ruthenium catalysts (0.74–3.06 wt%), based on ordered MCM-41 silica arrays on aluminosilicate halloysite nanotubes (HNTs), as well as HNT-based counterparts, were synthesized and tested in benzene hydrogenation. The TEM and low-temperature nitrogen adsorption/desorption analyses for the HNT core-shell silica composite-supported Ru catalysts proved the well ordered mesoporous silica structure of the MCM-41 type assembled on the outer surface of the HNTs and retained after Ru loading. According to calculations based on the TEM results, the higher the metal loading, the higher the sizes of the particles formed by both the MCM-41/HNT and the HNT-based catalysts. The latter ones had particles varying from 3.6 to 5.4 nm in diameter, located predominantly on the outer surface of the HNTs. For the MCM-41/HNT-supported ones, at higher metal content the particle size distribution became bimodal, and particles with more than 6.7–7.2 nm in diameter formed.

When the metal loading increased, the Ru nanoparticles both intercalated into the MCM-41/HNT porous system and agglomerated outside the mesopores.

For all samples based on HNT, as well as on MCM-41/HNT, the higher is metal loading was, the more the specific catalytic activity decreased. The higher hydrogenation activity of the catalysts with Ru loading < 1 wt% was explained by the formation of metal particles highly dispersed over the surface of the MCM-41/HNT composite. The Ru/MCM-41/HNT(1) catalyst with an average particle size of 3.4 ± 0.1 nm, being uniformly distributed over the surface of the mesoporous aluminosilicate support, was the most active in the hydrogenation of benzene to cyclohexane and the specific activity of 5594 $h^{-1}$ was achieved with $A_{spn}$ 0.39 mol·$h^{-1}$·$m^{-2}$. For catalysts based on the MCM-41/HNT composite with a higher Ru content, the decrease of the $A_{sp}$ and $A_{spn}$ parameters was close, which was caused by similar average NPs diameters, while the Ru content became the major factor influenced by catalytic activity. As for the Ru/HNT(2) and Ru/HNT(3), the $A_{spn}$ parameters were close, while the $A_{sp}$ was slightly reduced by 20%, confirming the negative enlargement effect of nanoparticles on a decrease in catalytic activity.

The MCM-41/HNT composite-supported Ru catalysts were found to be more stable under recycling due to hierarchical structure and a high specific surface area, resulting in the stabilization of the ruthenium nanoparticles within the porous system. As a result, the specific catalytic activity and final benzene conversion for Ru/MCM-41/HNT(1) after 3 cycles were comparable for those obtained with the fresh Ru/HNT(1).

These catalysts, based on a synergistically strong, new type material, consisting of synthetic mesoporous silica of MCM-41 type arrays on natural clay nanotubes, are safe, environmentally friendly materials and could be easily scaled up for industrial application.

**Author Contributions:** Conceptualization, A.V., A.G., V.N.; methodology, V.S., V.N., A.G.; software, A.V., A.P.; validation, A.V., A.P.; formal analysis, V.S., G.Z.; investigation, V.N., G.Z.; resources, V.V., A.G.; data curation, A.P., V.S., E.K.; writing—original draft preparation, V.N., G.Z., V.S.; writing—review and editing, A.V., A.G., A.P.; visualization, A.V., V.N.; supervision, V.A., A.G.; project administration, A.G.; funding acquisition A.G., A.P., A.V. All authors have read and agreed to the published version of the manuscript.

**Funding:** This work was financially supported by the Russian Science Foundation (Project No. 19-79-10016).

**Acknowledgments:** We thank Anna Stavitskaya and Ekaterina Smirnova (Gubkin University) for their input in this work. We thank Yuri Lvov from Louisiana Tech University (USA) for initiation this topic at Gubkin University and for collaboration.

**Conflicts of Interest:** The authors declare no conflict of interest.

## References

1. Alimohammadi, N.; Fathi, S. Selective benzene reduction from gasoline using catalytic hydrogenation reactions over zeolite Pd/13X. *React. Kinet. Mech. Catal.* **2019**, *128*, 949–964. [CrossRef]
2. Almering, M.; Rock, K.; Judzis, A. Reducing Benzene in Gasoline: Cost-Effective Solutions for the Reduction of Benzene in Gasoline, Particularly with Regard to Msat II Requirement. In Proceedings of the 106th NPRA Annual Meeting, San Diego, CA, USA, 9–11 March 2008.
3. Zhou, G.; Jiang, L.; Dong, Y.; Li, R.; He, D. Engineering the exposed facets and open-coordinated sites of brookite TiO2 to boost the loaded Ru nanoparticle efficiency in benzene selective hydrogenation. *Appl. Surf. Sci.* **2019**, *486*, 187–197. [CrossRef]
4. Foppa, L.; Dupont, J. Benzene partial hydrogenation: Advances and perspectives. *Chem. Soc. Rev.* **2015**, *44*, 1886–1897. [CrossRef] [PubMed]
5. Schwab, F.; Lucas, M.; Claus, P. Ruthenium-Catalyzed Selective Hydrogenation of Benzene to Cyclohexene in the Presence of an Ionic Liquid. *Angew. Chem. Int. Ed.* **2011**, *50*, 10453–10456. [CrossRef] [PubMed]
6. Weissermel, H.J.K. *Arpe Industrial Organic Chemistry*, 4th ed.; Wiley-VCH: Weinheim, Germany, 2003.
7. Tan, X.; Zhou, G.; Qiao, M.; Dou, R.; Sun, B.; Pei, Y.; Zong, B.; Fan, K. Partial hydrogenation of benzene to cyclohexene over novel Ru-B/MOF catalysts. *Acta Phys. Chim. Sin.* **2014**, *30*, 932–942.
8. Peng, Z.; Liu, X.; Li, S.; Li, Z.; Li, B.; Liu, Z.; Liu, S. Heterophase-structured nanocrystals as superior supports for Ru-based catalysts in selective hydrogenation of benzene. *Sci. Rep.* **2017**, *7*, 39847. [CrossRef] [PubMed]

9. Wang, Y.; Rong, Z.; Wang, Y.; Qu, J. Ruthenium nanoparticles loaded on functionalized graphene for liquid-phase hydrogenation of fine chemicals: Comparison with carbon nanotube. *J. Catal.* **2016**, *333*, 8–16. [CrossRef]
10. Hu, S.-C.; Chen, Y.-W. Partial Hydrogenation of Benzene to Cyclohexene on Ruthenium Catalysts Supported on La2O3–ZnO Binary Oxides. *Ind. Eng. Chem. Res.* **1997**, *36*, 5153–5159. [CrossRef]
11. Glotov, A.; Stytsenko, V.D.; Artemova, M.; Kotelev, M.; Ivanov, E.V.; Gushchin, P.; Vinokurov, V. Hydroconversion of Aromatic Hydrocarbons over Bimetallic Catalysts. *Catalysts* **2019**, *9*, 384. [CrossRef]
12. Goundani, K.; Papadopoulou, C.; Kordulis, C. Benzene elimination from reformate gasoline by high pressure hydrogenation in a fixed-bed reactor. *React. Kinet. Catal. Lett.* **2004**, *82*, 149–155. [CrossRef]
13. Sassykova, L. Development of Catalysts for the Hydrogenation of the Aromatic Ring in Gasolines. *Chem. Biochem. Eng. Q.* **2018**, *31*, 447–453. [CrossRef]
14. Wang, C.; Chen, X.; Chen, T.-M.; Wei, J.; Qin, S.-N.; Zheng, J.-F.; Zhang, H.; Tian, Z.-Q.; Li, J.-F. In-situ SHINERS Study of the Size and Composition Effect of Pt-based Nanocatalysts in Catalytic Hydrogenation. *Chem. Cat. Chem.* **2019**, *12*, 75–79. [CrossRef]
15. Glotov, A.; Stavitskaya, A.V.; Chudakov, Y.A.; Artemova, M.I.; Smirnova, E.M.; Demikhova, N.R.; Shabalina, T.N.; Gureev, A.A.; Vinokurov, V. Nanostructured Ruthenium Catalysts in Hydrogenation of Aromatic Compounds. *Pet. Chem.* **2018**, *58*, 1221–1226. [CrossRef]
16. Vinokurov, V.; Glotov, A.; Chudakov, Y.; Stavitskaya, A.; Ivanov, E.; Gushchin, P.; Zolotukhina, A.; Maximov, A.; Karakhanov, E.; Lvov, Y. Core/Shell Ruthenium–Halloysite Nanocatalysts for Hydrogenation of Phenol. *Ind. Eng. Chem. Res.* **2017**, *56*, 14043–14052. [CrossRef]
17. Mohammadian, Z.; Peyrovi, M.H.; Parsafard, N. Activity and Stability Evaluation of Nickel Supported over Al-Meso-Microporous Hybrids in Selective Hydrogenation of Benzene. *ChemistrySelect* **2019**, *4*, 11116–11120. [CrossRef]
18. Zhang, H.; Meng, Y.; Song, G.; Li, F. Effect of Hydrogen Spillover to the Hydrogenation of Benzene Over Pt/NaA Catalysts. *Synth. React. Inorg. Met. Chem.* **2015**, *46*, 940–944. [CrossRef]
19. Coumans, A.E.; Poduval, D.G.; Van Veen, J.R.; Hensen, E.J. The nature of the sulfur tolerance of amorphous silica-alumina supported NiMo(W) sulfide and Pt hydrogenation catalysts. *Appl. Catal. A Gen.* **2012**, *411*, 51–59. [CrossRef]
20. Pimerzin, A.A.; Roganov, A.A.; Verevkin, S.P.; Konnova, M.E.; Pilshchikov, V.A.; Pimerzin, A.A. Bifunctional catalysts with noble metals on composite $Al_2O_3$-SAPO-11 carrier and their comparison with CoMoS one in n-hexadecane hydroisomerization. *Catal. Today* **2019**, *329*, 71–81. [CrossRef]
21. Comyns, A.E. *Encyclopedic Dictionary of Named Processes in Chemical Technology*, 3rd ed.; CRC Press: Boca Raton, FL, USA, 2008; Volume 54, p. 3032.
22. Vutolkina, A.V.; Makhmutov, D.F.; Zanina, A.V.; Maximov, A.L.; Kopitsin, D.S.; Glotov, A.P.; Egazar'Yants, S.V.; Karakhanov, E.A. Hydroconversion of Thiophene Derivatives over Dispersed Ni–Mo Sulfide Catalysts. *Pet. Chem.* **2018**, *58*, 1227–1232. [CrossRef]
23. Liu, H.; Fang, R.; Li, Z.; Li, Y. Solventless hydrogenation of benzene to cyclohexane over a heterogeneous Ru–Pt bimetallic catalyst. *Chem. Eng. Sci.* **2015**, *122*, 350–359. [CrossRef]
24. Parsafard, N.; Rostamikia, T.; Parsafard, N. Competitive Hydrogenation of Benzene in Reformate Gasoline over Ni Supported on $SiO_2$, $SiO_2$–$Al_2O_3$, and $Al_2O_3$ Catalysts: Influence of Support Nature. *Energy Fuels* **2018**, *32*, 11432–11439. [CrossRef]
25. Roldugina, A.E.; Glotov, A.P.; Isakov, A.L.; Maksimov, A.L.; Vinokurov, V.A.; Karakhanov, E.A. Ruthenium Catalysts on Zsm-5/Mcm-41 Micro-Mesoporous Support for Hydrodeoxygenation of Guaiacol in the Presence of Water. *Russ. J. Appl. Chem.* **2019**, *92*, 1170–1178. [CrossRef]
26. Takasaki, M.; Motoyama, Y.; Higashi, K.; Yoon, S.H.; Mochida, I. Nagashima Ruthenium nanoparticles on nano-level-controlled carbon supports as highly effective catalysts for arene hydrogenation. *Chem. Asian J.* **2007**, *2*, 1524–1533. [CrossRef] [PubMed]
27. Duan, H.; Wang, D.; Kou, Y.; Li, Y. Rhodium-nickel bimetallic nanocatalysts: High performance of room-temperature hydrogenation. *Chem. Commun.* **2013**, *49*, 303–305. [CrossRef] [PubMed]
28. Zhu, L.; Zheng, L.; Du, K.; Fu, H.; Li, Y.; You, G.; Chen, B.H. An efficient and stable Ru–Ni/C nano-bimetallic catalyst with a comparatively low Ru loading for benzene hydrogenation under mild reaction conditions. *RSC Adv.* **2013**, *3*, 713–719. [CrossRef]

29. Glotov, A.; Stavitskaya, A.; Chudakov, Y.; Ivanov, E.V.; Huang, W.; Vinokurov, V.; Zolotukhina, A.; Maximov, A.; Karakhanov, E.; Lvov, Y. Mesoporous Metal Catalysts Templated on Clay Nanotubes. *Bull. Chem. Soc. Jpn.* **2019**, *92*, 61–69. [CrossRef]
30. Nedolivko, V.V.; Zasypalov, G.O.; Chudakov, Y.A.; Vutolkina, A.V.; Pimerzin, A.A.; Glotov, A.P. Effect of the ruthenium deposition method on the nanostructured catalyst activity in the deep hydrogenation of benzene. *Russ. Chem. Bull.* **2020**, *69*, 260–264. [CrossRef]
31. Zhu, L.; Sun, H.; Fu, H.; Zheng, J.; Zhang, N.; Li, Y.; Chen, B.H. Effect of ruthenium nickel bimetallic composition on the catalytic performance for benzene hydrogenation to cyclohexane. *Appl. Catal. A Gen.* **2015**, *499*, 124–132. [CrossRef]
32. Naranov, E.; Maximov, A.L. Selective conversion of aromatics into cis-isomers of naphthenes using Ru catalysts based on the supports of different nature. *Catal. Today* **2019**, *329*, 94–101. [CrossRef]
33. Glotov, A.; Artemova, M.I.; Demikhova, N.R.; Smirnova, E.M.; Ivanov, E.V.; Gushchin, P.A.; Egazar'Yants, S.V.; Vinokurov, V.A. A Study of Platinum Catalysts Based on Ordered Al–MCM-41 Aluminosilicate and Natural Halloysite Nanotubes in Xylene Isomerization. *Pet. Chem.* **2019**, *59*, 1226–1234. [CrossRef]
34. Karakhanov, E.A.; Glotov, A.; Nikiforova, A.G.; Vutolkina, A.V.; Ivanov, A.O.; Kardashev, S.V.; Maksimov, A.L.; Lysenko, S.V. Catalytic cracking additives based on mesoporous MCM-41 for sulfur removal. *Fuel Process. Technol.* **2016**, *153*, 50–57. [CrossRef]
35. Vutolkina, A.; Glotov, A.; Zanina, A.; Makhmutov, D.; Maximov, A.; Egazar'Yants, S.; Karakhanov, E.; Mahmutov, D.; Maksimov, A. Mesoporous Al-HMS and Al-MCM-41 supported Ni-Mo sulfide catalysts for HYD and HDS via in situ hydrogen generation through a WGSR. *Catal. Today* **2019**, *329*, 156–166. [CrossRef]
36. Xiao, Y.; Liao, H.; Yu, X.; Liu, X.; He, H.; Zhong, H.; Liang, M. Hydrophilicity modification of MCM-41 with polyethylene glycol and supported ruthenium for benzene hydrogenation to cyclohexene. *J. Porous Mater.* **2018**, *26*, 765–773. [CrossRef]
37. Martí, A.; Balmori, A.; Pontón, I.; Del Rio, A.M.; Sánchez-García, D. Functionalized Ordered Mesoporous Silicas (MCM-41): Synthesis and Applications in Catalysis. *Catalysts* **2018**, *8*, 617. [CrossRef]
38. Liang, J.; Liang, Z.; Zou, R.; Zhao, Y. Heterogeneous Catalysis in Zeolites, Mesoporous Silica, and Metal-Organic Frameworks. *Adv. Mater.* **2017**, *29*, 1701139. [CrossRef] [PubMed]
39. Glotov, A.; Levshakov, N.; Vutolkina, A.; Lysenko, S.; Karakhanov, E.; Vinokurov, V. Aluminosilicates supported La-containing sulfur reduction additives for FCC catalyst: Correlation between activity, support structure and acidity. *Catal. Today* **2019**, *329*, 135–141. [CrossRef]
40. Tian, P.; Blanchard, J.; Fajerwerg, K.; Breysse, M.; Vrinat, M.; Liu, Z. Preparation of Ru metal nanoparticles in mesoporous materials: Influence of sulfur on the hydrogenating activity. *Microporous Mesoporous Mater.* **2003**, *60*, 197–206. [CrossRef]
41. Galacho, C.; Carrott, M.R.; Carrott, P.J.M.; Galacho, P. Evaluation of the thermal and mechanical stability of Si-MCM-41 and Ti-MCM-41 synthesised at room temperature. *Microporous Mesoporous Mater.* **2008**, *108*, 283–293. [CrossRef]
42. Glotov, A.; Stavitskaya, A.; Novikov, A.; Semenov, A.; Ivanov, E.; Gushchin, P.; Darrat, Y.; Vinokurov, V.; Lvov, Y. Halloysite Based Core-Shell Nanosystems: Synthesis and Application. In *Nanomaterials from Clay Minerals*, 1st ed.; Wang, A., Wang, W., Eds.; Elsevier: Amsterdam, The Netherlands, 2019; pp. 203–256.
43. Lvov, Y.; Wang, W.; Zhang, L.; Fakhrullin, R. Halloysite Clay Nanotubes for Loading and Sustained Release of Functional Compounds. *Adv. Mater.* **2015**, *28*, 1227–1250. [CrossRef]
44. Sadjadi, S.; Lazzara, G.; Heravi, M.M.; Cavallaro, G. Pd supported on magnetic carbon coated halloysite as hydrogenation catalyst: Study of the contribution of carbon layer and magnetization to the catalytic activity. *Appl. Clay Sci.* **2019**, *182*, 105299. [CrossRef]
45. Vinokurov, V.; Stavitskaya, A.; Glotov, A.; Novikov, A.A.; Zolotukhina, A.V.; Kotelev, M.; Gushchin, P.A.; Ivanov, E.; Darrat, Y.; Lvov, Y. Nanoparticles Formed onto/into Halloysite Clay Tubules: Architectural Synthesis and Applications. *Chem. Rec.* **2018**, *18*, 858–867. [CrossRef] [PubMed]
46. Lvov, Y.; Panchal, A.; Fu, Y.; Fakhrullin, R.; Kryuchkova, M.; Batasheva, S.; Stavitskaya, A.V.; Glotov, A.; Vinokurov, V. Interfacial Self-Assembly in Halloysite Nanotube Composites. *Langmuir* **2019**, *35*, 8646–8657. [CrossRef] [PubMed]
47. Vinokurov, V.; Stavitskaya, A.; Glotov, A.; Ostudin, A.; Sosna, M.; Gushchin, P.; Darrat, Y.; Lvov, Y. Halloysite nanotube-based cobalt mesocatalysts for hydrogen production from sodium borohydride. *J. Solid State Chem.* **2018**, *268*, 182–189. [CrossRef]

48. Vinokurov, V.; Stavitskaya, A.; Chudakov, Y.A.; Glotov, A.; Ivanov, E.V.; Gushchin, P.A.; Lvov, Y.; Maximov, A.; Muradov, A.V.; Karakhanov, E.A. Core-shell nanoarchitecture: Schiff-base assisted synthesis of ruthenium in clay nanotubes. *Pure Appl. Chem.* **2018**, *90*, 825–832. [CrossRef]
49. Glotov, A.; Levshakov, N.; Stavitskaya, A.; Artemova, M.; Gushchin, P.; Ivanov, E.V.; Vinokurov, V.; Lvov, Y.; Levshakov, N. Templated self-assembly of ordered mesoporous silica on clay nanotubes. *Chem. Commun.* **2019**, *55*, 5507–5510. [CrossRef] [PubMed]
50. Liao, H.; Ouyang, D.; Zhang, J.; Xiao, Y.; Liu, P.; Hao, F.; You, K.; Luo, H. Influence of Preparation Conditions on the Structure of MCM-41 and Catalytic Performance of Ru/MCM-41 in Benzene Hydrogenation. *J. Chem. Res.* **2014**, *38*, 90–95. [CrossRef]
51. Vinokurov, V.; Stavitskaya, A.; Chudakov, Y.A.; Ivanov, E.V.; Shrestha, L.K.; Ariga, K.; Darrat, Y.A.; Lvov, Y. Formation of metal clusters in halloysite clay nanotubes. *Sci. Technol. Adv. Mater.* **2017**, *18*, 147–151. [CrossRef]
52. Koopman, P. Characterization of ruthenium catalysts as studied by temperature programmed reduction. *J. Catal.* **1981**, *69*, 172–179. [CrossRef]
53. Cao, S.; Tao, F.; Tang, Y.; Li, Y.; Yu, J. Size- and shape-dependent catalytic performances of oxidation and reduction reactions on nanocatalysts. *Chem. Soc. Rev.* **2016**, *45*, 4747–4765. [CrossRef]
54. Su, F.; Lv, L.; Lee, F.Y.; Liu, T.; Cooper, A.I.; Zhao, X.S. Thermally Reduced Ruthenium Nanoparticles as a Highly Active Heterogeneous Catalyst for Hydrogenation of Monoaromatics. *J. Am. Chem. Soc.* **2007**, *129*, 14213–14223. [CrossRef]
55. Nandanwar, S.U.; Chakraborty, M.; Mukhopadhyay, S.; Shenoy, K.T. Benzene hydrogenation over highly active monodisperse Ru/γ-Al$_2$O$_3$ nanocatalyst synthesized by (w/o) reverse microemulsion. *React. Kinet. Mech. Catal.* **2012**, *108*, 473–489. [CrossRef]
56. Wang, Y.; Rong, Z.; Wang, Y.; Zhang, P.; Wang, Y.; Qu, J. Ruthenium nanoparticles loaded on multiwalled carbon nanotubes for liquid-phase hydrogenation of fine chemicals: An exploration of confinement effect. *J. Catal.* **2015**, *329*, 95–106. [CrossRef]
57. Boricha, A.; Mody, H.; Bajaj, H.; Jasra, R.V. Hydrogenation of benzene over ruthenium-exchanged montmorillonite in the presence of thiophene. *Appl. Clay Sci.* **2006**, *31*, 120–125. [CrossRef]
58. Maximov, A.; Zolotukhina, A.; Kulikov, L.A.; Kardasheva, Y.; Karakhanov, E. Ruthenium catalysts based on mesoporous aromatic frameworks for the hydrogenation of arenes. *React. Kinet. Mech. Catal.* **2016**, *117*, 729–743. [CrossRef]
59. Chen, D.; Huang, M.; He, S.; He, S.; Ding, L.; Wang, Q.; Yu, S.; Miao, S. Ru-MOF enwrapped by montmorillonite for catalyzing benzene hydrogenation. *Appl. Clay Sci.* **2016**, *119*, 109–115. [CrossRef]
60. Sharma, S.K.; Sidhpuria, K.B.; Jasra, R.V. Ruthenium containing hydrotalcite as a heterogeneous catalyst for hydrogenation of benzene to cyclohexane. *J. Mol. Catal. A Chem.* **2011**, *335*, 65–70. [CrossRef]
61. Karakhanov, E.; Maximov, A.; Zolotukhina, A.; Vinokurov, V.; Ivanov, E.V.; Glotov, A. Manganese and Cobalt Doped Hierarchical Mesoporous Halloysite-Based Catalysts for Selective Oxidation of p-Xylene to Terephthalic Acid. *Catalysts* **2019**, *10*, 7. [CrossRef]

© 2020 by the authors. Licensee MDPI, Basel, Switzerland. This article is an open access article distributed under the terms and conditions of the Creative Commons Attribution (CC BY) license (http://creativecommons.org/licenses/by/4.0/).

Article

# Palladium Catalysts Based on Porous Aromatic Frameworks, Modified with Ethanolamino-Groups, for Hydrogenation of Alkynes, Alkenes and Dienes

**Leonid Kulikov [1], Maria Kalinina [1], Daria Makeeva [1], Anton Maximov [1,2], Yulia Kardasheva [1], Maria Terenina [1] and Eduard Karakhanov [1,*]**

1. Department of Chemistry, Lomonosov Moscow State University, 119234 Moscow, Russia; mailforleonid@mail.ru (L.K.); kalmary@yandex.ru (M.K.); d-makeeva95@yandex.ru (D.M.); max@ips.ac.ru (A.M.); yuskard@petrol.chem.msu.ru (Y.K.); tereninam@petrol.chem.msu.ru (M.T.)
2. Topchiev Institute of Petrochemical Synthesis, 119234 Moscow, Russia
* Correspondence: kar@petrol.chem.msu.ru; Tel.: +7-495-939-5377

Received: 28 August 2020; Accepted: 22 September 2020; Published: 24 September 2020

**Abstract:** The current work describes an attempt to synthesize hybrid materials combining porous aromatic frameworks (PAFs) and dendrimers and use them to obtain novel highly active and selective palladium catalysts. PAFs are carbon porous materials with rigid aromatic structure and high stability, and the dendrimers are macromolecules which can effectively stabilize metal nanoparticles and tune their activity in catalytic reactions. Two porous aromatic frameworks, PAF-20 and PAF-30, are modified step-by-step with diethanolamine and hydroxyl groups at the ends of which are replaced by new diethanolamine molecules. Then, palladium nanoparticles are applied to the synthesized materials. Properties of the obtained materials and catalysts are investigated using X-ray photoelectron spectroscopy, transmission electron microscopy, solid state nuclear magnetic resonance spectroscopy, low temperature $N_2$ adsorption and elemental analysis. The resulting catalysts are successfully applied as an efficient and recyclable catalyst for selective hydrogenation of alkynes to alkenes at very high (up to 90,000) substrate/Pd ratios.

**Keywords:** hydrogenation; heterogeneous catalysts; palladium; porous polymers; amines

## 1. Introduction

Palladium is one of the most widespread metals used in the selective hydrogenation of unsaturated compounds [1–3]. It is commonly used in a form of nanoparticles, for this process, whose high surface area provides a large number of available active sites per unit area. This leads to greater reactivity of the nanoparticles than with bulk palladium [4]. Changes in the size of particles, their morphology and distribution significantly affect their catalytic properties. However, unstabilized metal nanoparticles (NPs) are prone to aggregation, which leads to the formation of thermodynamically stable large particles with lower catalytic activity [5]. The main method to avoid aggregation of NPs is the use of encapsulating ligands and polymers or porous materials [6–11].

One of the most effective supports for metal nanoparticles are dendrimers—regular, three-dimensional, spherically symmetric macromolecules. Since dendrimers have interior voids of nanometer dimensions, they can hold nanoparticles of suitable sizes, which can fit into those voids [12]. Encapsulation of metals in a structure of dendrimers allows for control of the size of the nanoparticles and provides for their uniform distribution [13]. Strong coordination of the chelating nitrogen-containing groups with metal prevents NPs from leaching and agglomerating during reaction, thus providing outstanding stability. However, the use of dendrimer-stabilized nanoparticles is limited due to difficulties, connected to their separation from the reaction mixtures and time-consuming synthesis, which includes multistage purification [14].

These disadvantages may be overcome using one of two possible approaches. The first includes the covalent linking of dendrimer macromolecules using different bi- and tri-functional agents [15,16]. The second implies the attachment of dendrimers or dendrons to the surface of organic or inorganic insoluble supports. The main examples of such carriers are $SiO_2$ (both amorphous and mesoporous) [9,17] multiwall carbon nanotubes, or cross-linked polymer resins, such as polystyrenes [18], polythiophenes [19], or polyvinylpyridines [20]. The development of such hybrid materials is promising for the field of hydrogenation, due to their high activity and stability, immense selectivity, and easy recyclability.

Due to high values of specific surface area, developed porosity, and a variety of methods for the synthesis and modification of the structure, mesoporous materials and polymers have found wide application in the stabilization of nanoparticles [21–27]. One of these materials are porous aromatic frameworks (PAF)—polymers with a rigid structure consisting of aromatic rings connected to each other [28]. Their aromatic nature provides thermal stability and the possibility for facile introduction of functional groups, as well as additional stabilization of palladium nanoparticles. Modification of the polymer structure with various functional groups allows improvement of characteristics of the resulting catalyst significantly to increase its stability, to control activity, and selectivity [29–36]. All these properties make porous aromatic frameworks promising supports for metal nanoparticles. The incorporation of coordination electron-donating groups in the material is regarded as an especially efficient method to stabilize active catalytic species [27,36–46].

Materials with nitrogen and oxygen functionalities, such as amine and hydroxyl groups, have a high affinity for metal ions [47]. Porous organic polymers with a high nitrogen and oxygen content have a high potential for the uptake of metal ions or the immobilization of metal nanoparticles [18,48–52]. Such features, and their resulting applications in heterogeneous catalysis, are the main driving force behind the design and synthesis of porous organic frameworks with electron-donating groups. Here, we study palladium catalysts based on mesoporous aromatic frameworks modified with diethanolamine in a hydrogenation of different unsaturated compounds.

## 2. Results and Discussion

### 2.1. Synthesis and Characterization of Supports

Porous aromatic frameworks (PAF), PAF-20 and PAF-30, were synthesized according to the method described by Yuan Y. et al. [53] from tetrakis-(p-bromophenyl)methane and 1,4-phenylenediboronic acid or 4,4′-biphenyldiboronic acid. Consequently, PAF-20 and PAF-30 differed in the number of benzene rings between $sp^3$ carbon atoms in the nodes of the frameworks: PAF-20 had three benzene rings, and PAF-30 had four rings. Thus, PAF-20 typically had a smaller pore size, but a larger surface area compared to PAF-30 [24]. Subsequent modification of PAF structure with chloromethyl groups and diethanolamino groups was performed using methods by Gangadharan. D. et al. [54] and Lu.W. et al. [55] (Scheme 1). To determine the structural features of the structure of the obtained materials, they were studied using solid state nuclear magnetic resonance (NMR) spectroscopy, low-temperature $N_2$ adsorption and elemental analysis.

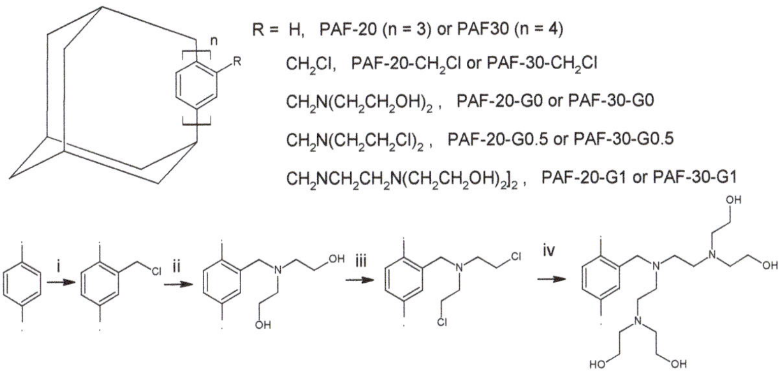

**Scheme 1.** Modification of PAF-20 and PAF-30. Reagents: (i) CH$_2$O, HCl, P$_2$O$_5$, AcOH, (ii) diethanolamine, dioxane, (iii) SOCl$_2$, dioxane, (iv) NH(EtOH)$_2$, dioxane.

Figure 1 shows the solid-state cross polarization-magic angle spinning (CP-MAS) $^{13}$C NMR spectra for PAF-based materials. All spectra contained typical for PAF signals in the region of 125–145 ppm for sp$^2$-hybridized carbon atoms of aromatic rings and, at 63 ppm, corresponded to sp$^3$-hybridized carbon atoms in the center of tetraphenylmethane units. After modification with chloromethyl groups, a new signal belonging to the -CH$_2$Cl group appeared at 44 ppm, which was consistent with published data [56,57]. Replacing the chlorine atom in -CH$_2$Cl with an ethanolamine group reduced the intensity of this peak due to screening of the carbon atom by branched ethanolamine groups and shifted its position from 44 ppm to 49–50 ppm. A similar dependence of the signal displacement upon the addition of various amines was observed during the earlier works [54,56]. Also, diethanolamino-modified materials PAF-20-G0, PAF-20-G1, PAF-30-G0 and PAF-30-G1 contained new signals at 56 and 58 ppm, which were assigned to carbon atoms of the ethanolamine groups.

Porous properties of PAFs were defined using low-temperature N$_2$ adsorption. Starting materials PAF-20 and PAF-30 had 578 m$^2$/g and 506 m$^2$/g Brunauer–Emmett–Teller (BET) surface area, respectively (Table 1). The surface area of polymers decreased upon modification: after chloromethylation it was 472 and 436 m$^2$/g for PAF-20-CH$_2$Cl and PAF-30-CH$_2$Cl, and after treatment with diethanolamine it was 64 and 31 m$^2$/g, respectively. Transition from G0 to G1 generation resulted in an even greater reduction of surface area, which was explained by the blocking of pores by bulky functional groups.

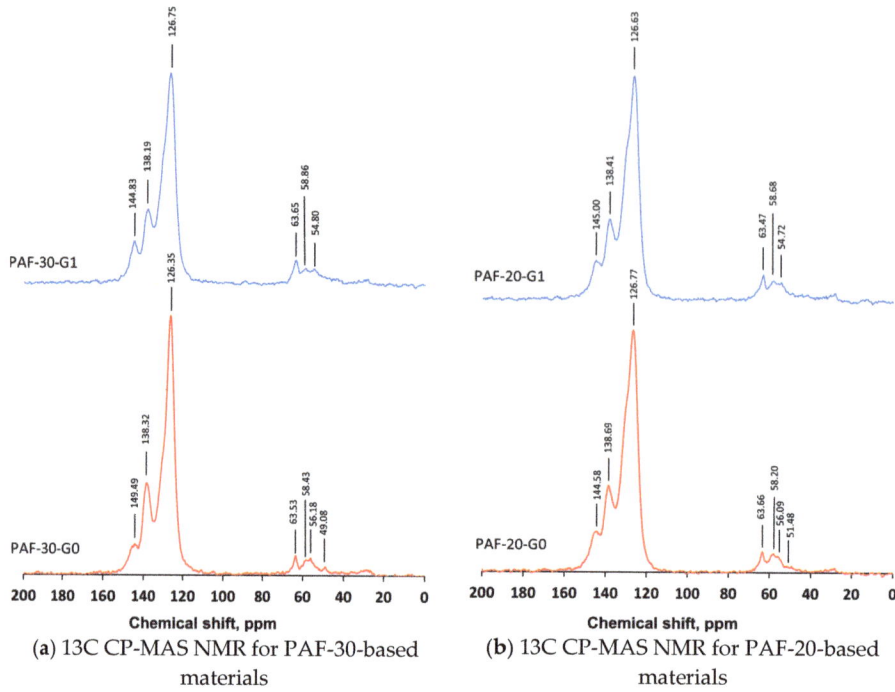

**Figure 1.** Solid-state 13C cross polarization-magic angle spinning nuclear magnetic resonance (CP-MAS NMR) spectra of obtained materials.

**Table 1.** Results of low-temperature nitrogen adsorption-desorption for obtained materials.

| Samples | Materials Based on PAF-20 | | Materials Based on PAF-30 | |
|---|---|---|---|---|
| | $S_{BET}$, m$^2$/g | Total Pore Volume (BJH), cm$^3$/g | $S_{BET}$, m$^2$/g | Total Pore Volume (BJH), cm$^3$/g |
| PAF | 579 | 0.316 | 506 | 0.311 |
| PAF-CH$_2$Cl | 472 | 0.264 | 436 | 0.262 |
| PAF-G0 | 29 | 0.026 | 61 | 0.054 |
| PAF-G1 | 5 | 0.001 | 38 | 0.007 |

The adsorption isotherms (Figure 2) of the samples PAF-20 and PAF-30, as well as their chloromethylated derivatives, exhibited a sharp absorption of N$_2$ at a low relative pressure (p/p$_0$ <0.05), which indicated the developed microporous structure of these materials. The presence of a hysteresis loop, and the fact that the nitrogen sorption curve gradually rose without the appearance of a plateau, may indicate the presence of mesopores in the polymer structure [58]. Conversely, this species may be due to diffusion restrictions or polymer swelling [59]. Chloromethylation did not change the character of nitrogen adsorption, which may indicate a uniform distribution of -CH$_2$Cl groups over the volume of the carrier and a slight change in the pore size.

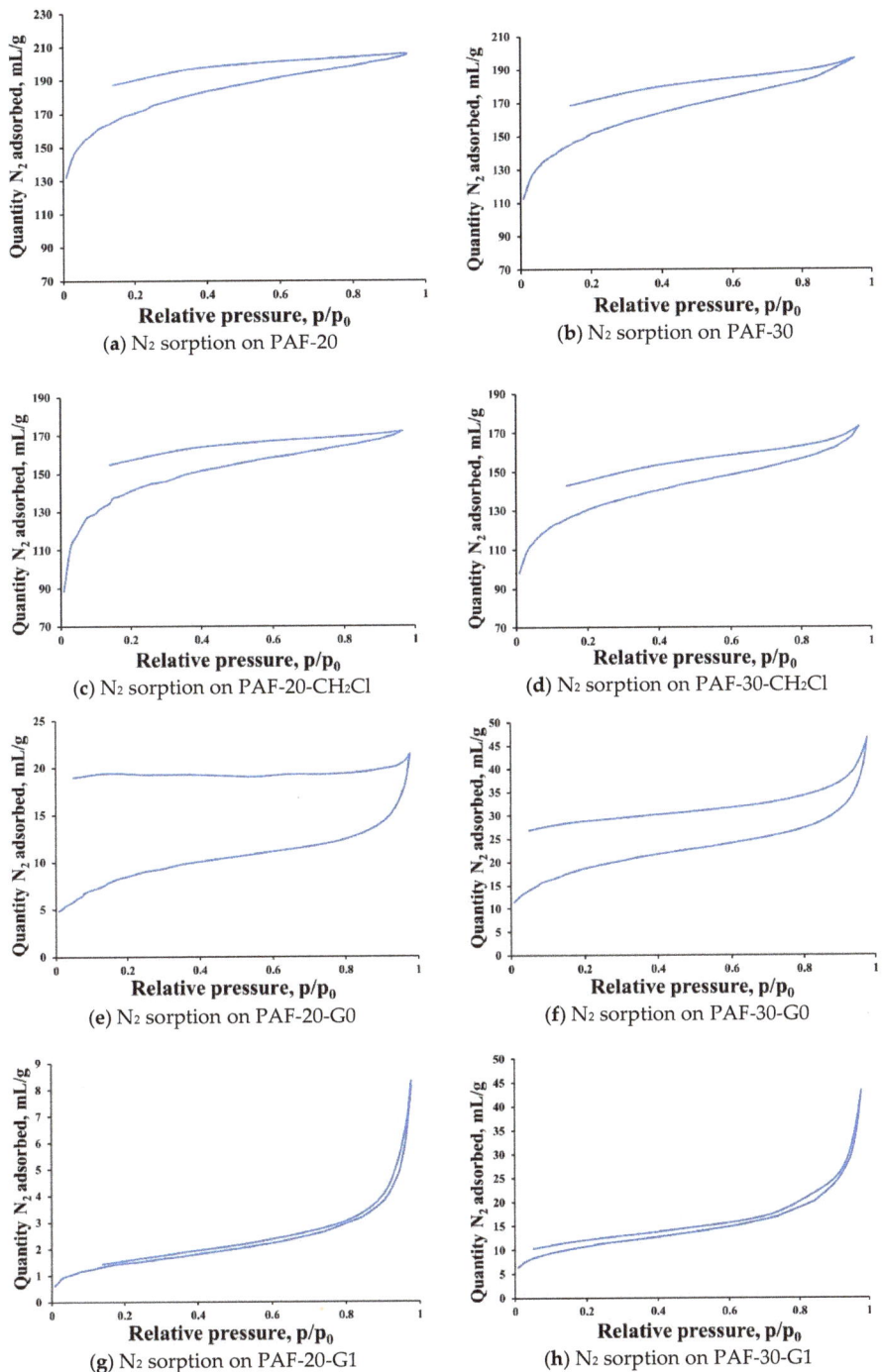

**Figure 2.** Low-temperature $N_2$ adsorption on PAF-based materials.

Treatment of chloromethylated polymers with diethanolamine led to significant change in the character of nitrogen adsorption: there was no longer a sharp rise in the adsorption curve and, in the case of PAF-20-G0, there was practically no desorption of nitrogen from the pores. This shape of the curves can be associated with strong blocking of the pores by the diethanolamino-groups, which was more pronounced for PAF-20-G0 due to its smaller pore size. Further modification of materials up to generation G1 led to a further reduction in pore size: adsorption curves for PAF-20-G1 and PAF-30-G1 were typical for non-porous materials [58]. Thus, we can conclude for PAF-20-G0 and PAF-30-G0 materials, only a small fraction of the pores were available and. in the case of PAF-20-G1 and PAF-30-G1, the only pores available were in the immediate vicinity of the outer surface of the catalyst grain. Also, the decrease in surface area and porosity after grafting of the surface of porous materials is a well-known fact. Thus, after modification of PAFs with different polyamines [58], the surface area and free volume of pores decreased dramatically from 4023 $m^2$/g for PPN-6 material to 555 $m^2$/g for tris-(2-aminoethyl)amine-modified material. However, we should note that in the case of the modification of porous aromatic frameworks with diethanolamine led to too high a decrease in porosity.

Elemental analysis also suggests successful functionalization of aromatic rings with different functional groups (Table 2). The averaged content of chlorine in the materials PAF-20-$CH_2$Cl and PAF-30-$CH_2$Cl was 3%—to be exact, about 7% of the benzene rings were modified. After treatment of chloromethylated polymers with diethanolamine, it decreased to 0.9%–1.2%, whereas nitrogen content in the resulting samples was 1.7%–1.8%. The presence of chlorine in these samples may indicate an incomplete modification: most likely, chloromethyl groups deep into the catalyst grain did not react with diethanolamine. Regarding the materials of the G1 series, the content of chlorine was even higher, and the nitrogen content increased only slightly. It indicates the occurrence of the substitution reaction of the hydroxyl groups for chlorine atoms during the treatment of materials of the G0 series with thionyl chloride, on the one hand. The completeness of this reaction was even lower than in the synthesis of materials of the G0 series, on the other hand.

Table 2. Elemental analysis of synthesized materials.

| Material | Element Content, Mass. % | |
|---|---|---|
| | Cl | N |
| PAF-20-$CH_2$Cl | 3.08% | - |
| PAF-30-$CH_2$Cl | 3.00% | - |
| PAF-20-G0 | 0.91% | 1.84% |
| PAF-30-G0 | 1.19% | 1.68% |
| PAF-20-G1 | 2.43% | 1.88% |
| PAF-30-G1 | 3.30% | 2.16% |

## 2.2. Characterization of Palladium Catalysts

Catalysts based on porous aromatic frameworks modified with the ethanolamine groups Pd-PAF-20-G0, Pd-PAF-20-G1, Pd-PAF-30-G0 and Pd-PAF-30-G1 were obtained by immobilizing palladium nanoparticles into pores of supports. The palladium content was determined by atomic absorption spectroscopy (AAS) (Table 3).

Pd-PAF-20-G0 and Pd-PAF-30-G0 catalysts contained well-dispersed small particles with a relatively narrow size distribution (2–2.5 nm) based on the transmission electron microscopy (TEM) microphotographs (Figure 3). This confirms the successful introduction of nanoparticles into the modified pores of materials.

Table 3. Palladium content by atomic absorption spectroscopy in synthesized catalysts.

| Material | Pd-PAF-20-G0 | Pd-PAF-30-G0 | Pd-PAF-20-G1 | Pd-PAF-30-G1 |
|---|---|---|---|---|
| Pd, mass % | 2.4 | 1.0 | 0.6 | 1.8 |

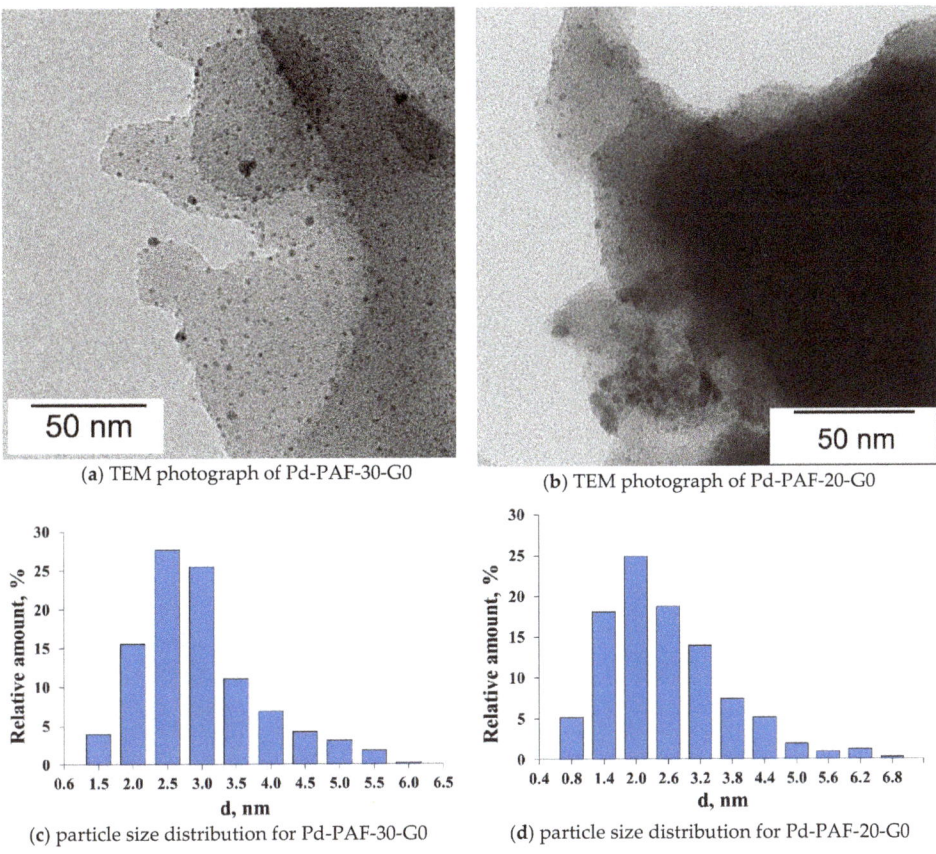

**Figure 3.** Transmission electron microscopy microphotographs and particle size distribution for Pd-PAF-20-G0 and Pd-PAF-30-G0.

To contrast, the G1 series catalysts contained only a small number of palladium particles, which were larger (average size 7–8 nm) and the size distribution curves were wider (Figure 4). It is seen in the micrographs that there are few nanoparticles and agglomerates are observed on the surface. This fact may be due to the blocking of the pores by the diethanolamine groups, which interfere with the diffusion of palladium ions inside the porous structure.

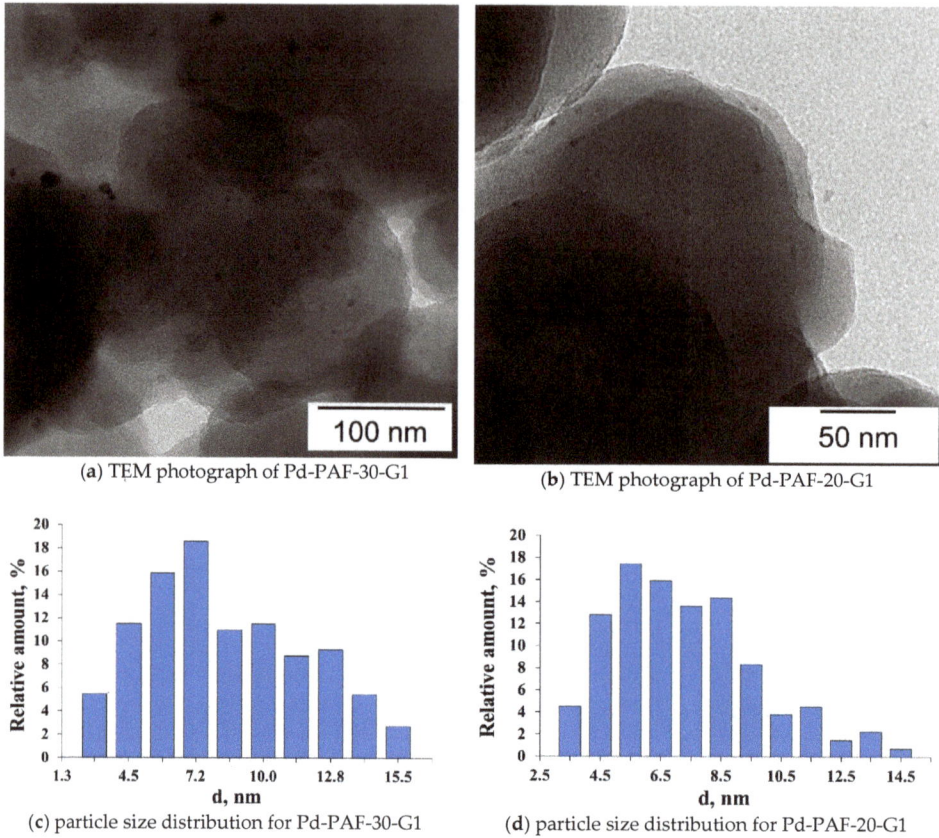

**Figure 4.** TEM microphotographs and particle size distribution for Pd-PAF-20-G1 and Pd-PAF-30-G1.

According to the X-ray photoelectron spectroscopy (XPS) data (Table 4), the nitrogen content was higher in the materials of the G1 series, which proves the process of modification of ethanolamine groups during the synthesis of materials PAF-20-G1 and PAF-30-G0. Concurrently, the chlorine content in the materials with modification of the G1 type remained higher than in the G0 series. That demonstrates the incomplete reaction of the substitution of chlorine for diethanolamine at the last stage of synthesis. The palladium content was higher in materials of the G0 series, which may be due to better immobilization of the palladium particles in the pores of the support.

**Table 4.** Components of the XPS spectra.

| Catalysts | C | O | N | Pd | Cl |
|---|---|---|---|---|---|
| Pd-PAF-20-G0 | 82.2 at.% | 13.3 at.% | 1.5 at.% | 2.8 at.% | 0.1 at.% |
| Pd-PAF-30-G0 | 85.6 at.% | 11.2 at.% | 1.7 at.% | 1.3 at.% | 0.2 at.% |
| Pd-PAF-20-G1 | 87.9 at.% | 6.8 at.% | 3.4 at.% | 0.3 at.% | 0.6 at.% |
| Pd-PAF-30-G1 | 87.0 at.% | 9.2 at.% | 3.0 at.% | 0.3 at.% | 0.5 at.% |

All XPS spectra demonstrated two sets of peaks related to reduced ($Pd^0$) and oxidized (PdOx) palladium forms (Figure 5). The presence of oxides in the spectra of samples could be due to incomplete reduction of the initial palladium or oxidation of the nanoparticles with atmospheric oxygen. The binding energies of $Pd^0\ 3d_{5/2}$ and $Pd^0\ 3d_{3/2}$ for Pd-PAF-20-G0 and Pd-PAF-30-G0 catalysts were higher than that for free Pd$^0$ (335.6 and 341.1 eV, respectively). The changing of energy values indicates the presence of a coordination effect between nitrogen or oxygen atoms in the supports (Table 5). The same binding energies for the materials Pd-PAF-20-G1 and Pd-PAF-30-G1 were practically no different from the binding energies for free palladium. These results confirm the assumption that there is no coordination between palladium nanoparticles and the diethanolamine groups. Nanoparticles in these catalysts are either in unmodified pores or on the surface of the support.

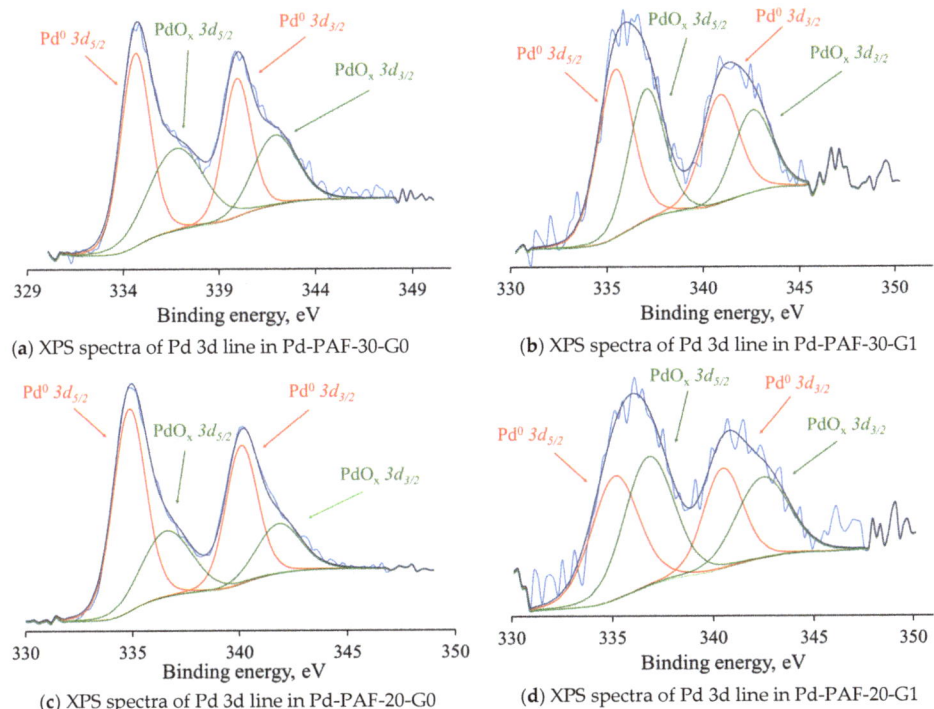

(a) XPS spectra of Pd 3d line in Pd-PAF-30-G0
(b) XPS spectra of Pd 3d line in Pd-PAF-30-G1
(c) XPS spectra of Pd 3d line in Pd-PAF-20-G0
(d) XPS spectra of Pd 3d line in Pd-PAF-20-G1

**Figure 5.** X-ray photoelectron spectroscopy (XPS) measurements for obtained palladium catalysts.

**Table 5.** Peak parameters for XPS spectra of obtained palladium catalysts.

| Catalyst | Parameter | Pd$^0$ | PdO$_x$ |
|---|---|---|---|
| Pd-PAF-20-G0 | Binding energy, eV | Pd 3d$_{5/2}$ 334.85 eV<br>Pd 3d$_{3/2}$ 340.10 eV | Pd 3d$_{5/2}$ 336.58 eV<br>Pd 3d$_{3/2}$ 341.82 eV |
| | Content, % | 67 | 33 |
| Pd-PAF-30-G0 | Binding energy, eV | Pd 3d$_{5/2}$ 334.65 eV<br>Pd 3d$_{3/2}$ 339.96 eV | Pd 3d$_{5/2}$ 336.75 eV<br>Pd 3d$_{3/2}$ 341.91 eV |
| | Content, % | 56 | 44 |
| Pd-PAF-20-G1 | Binding energy, eV | Pd 3d$_{5/2}$ 335.11 eV<br>Pd 3d$_{3/2}$ 340.47 eV | Pd 3d$_{5/2}$ 336.8 eV<br>Pd 3d$_{3/2}$ 342.43 eV |
| | Content, % | 48 | 51 |
| Pd-PAF-30-G1 | Binding energy, eV | Pd 3d$_{5/2}$ 335.45 eV<br>Pd 3d$_{3/2}$ 340.89 eV | Pd 3d$_{5/2}$ 337.07 eV<br>Pd 3d$_{3/2}$ 342.56 eV |
| | Content, % | 56 | 44 |

## 2.3. Catalytic Activity

The synthesized catalysts were examined in the hydrogenation of various $C_6$ and $C_8$ unsaturated compounds (Table 6). The G0 series catalysts were more active than the G1 series catalysts, and catalysts based on the materials of the PAF-30 type were more active than those of the PAF-20 type. Thus, the Pd-PAF-30-G0 catalyst showed the highest activity in the hydrogenation of linear alkynes and alkenes: its specific activity (Table 7) was more than 300,000 mol $_{Sub}$ × mol $_{Me}^{-1}$ × h$^{-1}$ in the case of hexyne-1, hexene-1, octyne-1, and octene-1 (Table 7). Regarding 2,5-dimethyl-2,4-hexadiene, it was about 295,000 mol $_{Sub}$ × mol $_{Me}^{-1}$ × h$^{-1}$ and, for more bulky substrates—phenylacetylene and styrene—it was much lower (116,000 and 79,000 mol $_{Sub}$ × mol $_{Me}^{-1}$ × h$^{-1}$, respectively), which may be due to conjugation between the benzene ring and double and triple bonds in these substrates.

Table 6. Hydrogenation of unsaturated hydrocarbons on palladium catalysts.

| Substrate | Reaction Products | Product Yield, % | | | |
|---|---|---|---|---|---|
| | | Pd-PAF-20-G0 | Pd-PAF-30-G0 | Pd-PAF-20-G1 | Pd-PAF-30-G1 |
| Hexyne-1 | Hexene-1 | 85 | 94 | 9 | 35 |
| | Hexane | 4 | 6 | - | - |
| Hexene-1 | Hexane | 34 | 100 | <1 | <1 |
| Cyclohexene | Cyclohexane | 11 | 12 | - | - |
| 1,3-cyclohexadiene | Cyclohexadiene | 7 | 9 | - | - |
| Octyne-1 | Octene-1 | 6 | 99 | - | - |
| Octyne-4 | Octene-4 | 3 | 4 | - | - |
| Octene-1 | Octane | 7 | 99 | <1 | 1 |
| | Isomerization products | 85 | <1 | 5 | 5 |
| 2,5-dimethyl-2,4-hexadiene | 2,5-dimethyl-3-hexene | 8 | 5 | <1 | <1 |
| | 2,5-dimethylhexane | <1 | 5 | <1 | 1 |
| | 2,5-dimethyl-2-hexene | 18 | 82 | - | - |
| Phenylacetylene | Styrene | 21 | 37 | - | - |
| Styrene | Ethylbenzene | 10 | 26 | - | - |
| 4-methoxystyrene | 4-methoxyethylbenzene | 3 | 4 | - | - |

Reaction conditions: 1 mg of the catalyst; substrate:metal = 22,500:1 (Pd-PAF-20-G0), 54,000:1 (Pd-PAF-20-G1), 90,000:1 (Pd-PAF-30-G0), 30,000:1 (Pd-PAF-30-G1), 80 °C, 1.0 MPa $H_2$, 30 min.

Table 7. Specific activity of synthesized catalysts.

| Substrate | Pd-PAF-20-G0 | Pd-PAF-30-G0 | Pd-PAF-20-G1 | Pd-PAF-30-G1 |
|---|---|---|---|---|
| Hexyne-1 | 94,600 | 323,400 | 100,700 | 189,800 |
| Hexene-1 | 34,600 | 305,100 | - | - |
| Cyclohexene | 11,200 | 36,600 | - | - |
| 1,3-cyclohexadiene | 7100 | 27,400 | - | - |
| Octyne-1 | 7100 | 302,000 | - | - |
| Octyne-4 | 5600 | 12,200 | - | - |
| Octene-1 | 7100 | 305,100 | - | - |
| 2,5-dimethyl-2,4-hexadiene | 27,500 | 294,900 | - | - |
| Phenylacetylene | 22,400 | 115,900 | - | - |
| Styrene | 10,200 | 79,300 | - | - |
| 4-methoxystyrene | 5600 | 12,200 | - | - |

Pd-PAF-20-G0 possessed lesser hydrogenation activity: high conversion was achieved only for hexyne-1, whereas yields of hydrogenation products for other substrates were much lower. This could be due to the smaller pore size in PAF-20 and, as a consequence, lower diffusion of substrates to palladium nanoparticles [33]. Catalysts Pd-PAF-20-G1 and Pd-PAF-30-G1 showed less activity. This may be due to the fact that further modification of the material with diethanolamine groups leads to a significant decrease in pore size, restricting the penetration of substrates, and the leaching of metal particles from the surface.

The stability of Pd-PAF-20-G0 and Pd-PAF-30-G0 catalysts was tested in recycle experiments regarding the hydrogenation of hexene-1. Both catalysts remained active for five reuse cycles. A slight reduction in conversions was observed only in the first two repetitions (Table 8). This fact is associated with the leaching of particles from the surface of the catalyst, as is seen in TEM micrographs, as well as in losses during catalytic experiments. A slight increase in the average particle size from 2.5 to 2.6 nm for Pd-PAF-30-G0, and from 2 to 2.9 nm for Pd-PAF-20-G0, can be noted; however, the distribution of the nanoparticles in the materials remained uniform (Figure 6). The washing out of nanoparticles from the surface also is confirmed by XPS data, according to which, after five reuse cycles, the amount of surface palladium decreased from 2.8 at.% to 2.1 at.% for Pd-PAF-20-G0, and from 1.3 at.% to 1.0 at.% for Pd PAF 30 G0 (Table 9).

Table 8. Results of recyclability tests.

| Catalyst | Product Yield, % | | | | |
|---|---|---|---|---|---|
| | Cycle 1 | Cycle 2 | Cycle 3 | Cycle 4 | Cycle 5 |
| Pd-PAF-20-G0 | 34 | 28 | 16 | 16 | 17 |
| Pd-PAF-30-G0 | 99 | 98 | 86 | 84 | 83 |

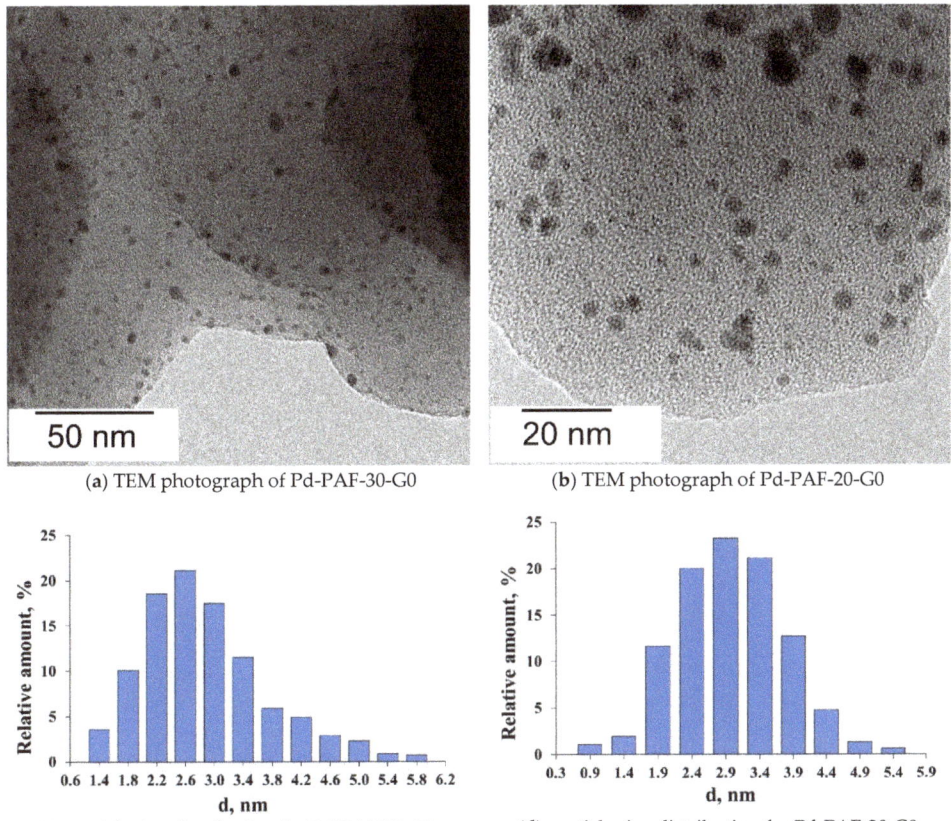

(a) TEM photograph of Pd-PAF-30-G0
(b) TEM photograph of Pd-PAF-20-G0
(c) particle size distribution for Pd-PAF-30-G0
(d) particle size distribution for Pd-PAF-20-G0

Figure 6. TEM microphotographs and particle size distribution for Pd-PAF-20-G0 and Pd-PAF-30-G0 after five runs of the recycle process.

Table 9. Components of the XPS spectra after five runs of the recycle process.

| Sample | | C | O | N | Cl | Pd |
|---|---|---|---|---|---|---|
| Pd-PAF-20-G0 | Before reaction | 82.2 at.% | 13.3 at.% | 1.5 at.% | 0.1 at.% | 2.8 at.% |
| | After 5 runs | 79.4 at.% | 16.1 at.% | 2.1 at.% | 0.2 at.% | 2.1 at.% |
| Pd-PAF-30-G0 | Before reaction | 85.6 at.% | 11.2 at.% | 1.7 at.% | 0.2 at.% | 1.3 at.% |
| | After 5 runs | 85.6 at.% | 12.3 at.% | 0.9 at.% | 0.2 at.% | 1.0 at.% |

## 3. Materials and Methods

### 3.1. Used Reagents

The following reagents were used in the work: benzene (IREA 2000, Moscow, Russia, Purum p.a.); styrene (Aldrich, St. Louis, MO, USA, ≥99%); phenylacetylene (Aldrich, Shanghai, China, 98%); 2,5-dimethyl-2,4-hexadiene (Aldrich, St. Louis, MO, USA, 98%); methanol (Acros Organics, Morris Plains, NJ, USA ); ethanol (IREA 2000, Moscow, Russia, Purum p.a.); chloroform (Ecos-1, Moscow, Russia, Purum), octene-1 (Aldrich, St. Louis, MO, USA, 98%); octyne-1 (abcr, Karlsruhe, Germany, 98%); 2,5-dimethylhexadiene-2,4 (Aldrich, St. Louis, MO, USA, 96%), hexene-1 (Aldrich, St. Louis, MO, USA, 98%), hexyne-1 (Aldrich, St. Louis, MO, USA, 99%), 1,3–cyclohexadiene (Acros Organics, Morris Plains, NJ, USA), paraformaldehyde (Sigma–Aldrich, St. Louis, MO, USA, 95%), hydrochloric acid (Sigma-tech, Moscow region, Russia, high-purity grade), phosphorus oxide(V) (Khimmed, Moscow, high-purity grade), acetic acid (Ruskhim, Moscow, Russia, high-purity grade), 1,4-dioxane (Ruskhim, Moscow, Russia, high-purity grade), diethanolamine (Sigma–Aldrich, St. Louis, MO, USA, 98%), potassium iodide (Reakhim, Staraya Kupavna, Moscow region, Russia, high-purity grade), acetone (Ekros, Saint-Peterburg, Russia, high-purity grade) thionyl chloride (Sigma–Aldrich, St. Louis, MO, USA, 97%), nitric acid (Component–Reaktiv, Moscow, Russia, high-purity grade), potassium carbonate (Component-reactive, Moscow, Russia, pure), sodium borohydride (Aldrich, St. Louis, MO, USA, 98%).

PAFs ware prepared according to published literature procedures [53]. The modification techniques for PAF-20 and PAF-30 are similar. Below are the methods for modifying the PAF-20 material.

### 3.2. Synthesis of PAF-20-CH$_2$Cl

Chloromethylation was carried out according to the method modified from a previous work [55]. Paraformaldehyde (1 g) and hydrochloric acid (20 mL) were placed in a round-bottomed flask equipped with a stirrer and a reflux condenser. After dissolving of paraformaldehyde, phosphorus pentoxide (4 g) and glacial acetic acid (6 mL) were carefully added to the mixture. Subsequently, PAF-20 (200 mg) was placed in the flask and stirred at 90 °C for 3 days. The resulting solid was collected using filtration, washed 3 times with water (100 mL) and ethanol (100 mL), and dried in vacuo to produce PAF-20-CH$_2$Cl as a yellow powder.

### 3.3. Synthesis of PAF-20-G0

Modification of PAF-20-CH$_2$Cl was performed according to the procedure adapted from a previous work [54]. PAF-20-CH$_2$Cl (100 mg), dioxane (6 mL), diethanolamine (0.4 mL) and catalytic amounts of potassium iodide (10 mg) were mixed in a round-bottomed flask and stirred at 70 °C for 3 days. The solid was collected using filtration, washed with acetone (3 × 50 mL), and dried in vacuo. The obtained material, PAF-20-CH$_2$N(CH$_2$CH$_2$OH)$_2$, was called PAF-20-G0 as an analogy with the numbering of the dendrimer generations.

### 3.4. Synthesis of PAF-20-G1

PAF-20-G0 (100 mg) was mixed with 1,4-dioxane (6 mL) in a round-bottomed flask, and then thionyl chloride (1.0 mL) was added dropwise to the suspension. The resulting mixture was stirred at 80 °C for 24 h, then the solid was collected using filtration and washed with ethanol (3 × 50 mL), a 2M solution of potassium carbonate (3 × 50 mL), water (3 × 50 mL) and ethanol (3 × 50 mL), and dried in vacuo. Then, the resulting material PAF-20-CH$_2$N(CH$_2$CH$_2$Cl)$_2$, called PAF-20-G0.5, was placed in a round-bottomed flask with dioxane (6 mL), diethanolamine (0.8 mL) and potassium iodide (10 mg). The suspension was stirred at 70 °C for 3 days. The resulting material was filtered off, washed with acetone (3 × 50 mL) and dried in vacuo.

### 3.5. Synthesis of Catalyst Pd-PAF-G0 and Pd-PAF-G1

Immobilization of palladium particles into the pores of PAFs was performed by the method described by Karakhanov, E.A. et al. [16]. The synthesis procedure described below was used to prepare the catalysts Pd-PAF-20-G0, Pd-PAF-20-G1, Pd-PAF-30-G0 and Pd-PAF-30-G1. The procedure for the synthesis of catalyst Pd-PAF-20-G0 is an example.

Palladium acetate (4.26 mg) was dissolved in chloroform (6 mL) in a round-bottomed flask, equipped with a magnetic stirrer and a reflux condenser. Then, PAF-20-G0 (100 mg) was added to the resulting solution. The suspension was stirred at 60 °C for 24 h, and then a solution of sodium borohydride (7 mg) in a water-ethanol mixture (0.5 mL:0.5 mL) was added. The reaction mixture darkened and gas evolution was observed. Subsequently, the flask was closed with a stopper and left for 12 h with stirring. After the reaction, the resulting substance was collected by centrifugation, washed with water (3 × 50 mL) and ethanol (3 × 50 mL) to remove sodium tetraborate, the precipitate was isolated by centrifugation and dried in air.

### 3.6. Catalytic Experiments

The calculated amounts of substrates and catalyst (1 mg) were placed in a glass tube equipped with a magnetic stirrer. The tube was placed in a steel autoclave, then it was sealed and pressurized with hydrogen at a pressure of 10 atm. Reactions were carried out at 80 °C for 30 min. After completion of the reaction, the autoclave was cooled to room temperature and depressurized. Reaction products were analyzed by gas chromatography. Specific activity of the catalysts was calculated using following equation:

$$A = \frac{Conv * (sub/Me)}{D * t} \quad (1)$$

where A is the specific activity, Conv is the conversion of the substrate, Sub/Me is the substrate to metal proportion, D is the metal dispersion, and t is the reaction time.

### 3.7. Characterization

#### 3.7.1. Low Temperature Nitrogen Adsorption

Nitrogen desorption/desorption isotherms were recorded at 77 K with a Micromeritics Gemini VII 2390 instrument (Micromeritics, Norcross, GA, United States). All samples were degassed at 110 °C for 6 h before measurement. The surface area ($S_{BET}$) was calculated using the Brunauer–Emmett–Teller (BET) method based on adsorption data in the relative pressure range $P/P_0 = 0.05$–$0.2$. The total pore volume (Vtot) was determined by the amount of nitrogen adsorbed at a relative pressure of $P/P_0 = 0.995$.

#### 3.7.2. Transmission Electron Microscopy (TEM)

TEM analysis was carried out on a JEOL JEM-2100/Cs/GIF microscope (JEOL, Tokyo, Japan) with a 0.19 nm lattice fringe resolution and an accelerating voltage of 200 kV. The processing of the micrographs and the calculation of the average particle size were conducted using the ImageJ software program.

#### 3.7.3. X-ray Photoelectron Spectroscopy (XPS)

XPS studies were performed on a VersaProbeII, ULVAC-PHI (ULVAC-PHI, Inc., Kanagawa, Japan) instrument using excitation with Al Kα X-ray radiation at 1486.6 eV. The calibration of photoelectron peaks was based on the Au 4f line with a binding energy of 84 eV and on the Cu2p3/2 line (932.6 eV). The transmission energy of the energy analyzer was 117.4 eV (survey scans) and 23.5 eV (individual lines).

3.7.4. Gas-Liquid Chromatography

Analysis of the reaction mixture was carried out on a Agilent 6890 G1530A chromatograph (Hewlett Packard, Santa-Clara, CA, United States) equipped with a flame-ionization detector and a HP-1 column (50 m × 0.32 mm × 1.05 μm, 100% dimethylpolysiloxane grafted phase). Helium was a carrier gas; the analysis was carried out in constant pressure mode (1.53 bar). Chromatograms were recorded and analyzed on a computer using the HP ChemStation Rev.A.06.01 (403) software.

3.7.5. Atomic Absorption Spectroscopy

The Pd content in the catalysts was determined via atomic absorption spectroscopy (AAS) on an iCE 3000 Series AA spectrometer (Thermo Scientific, Santa-Clara, CA, United States) with flame atomization. The data were processed using the SOLAAR software.

## 4. Conclusions

To conclude, we developed active and selective catalysts based on hybrid materials. It was shown that introduction of ethanolamine groups allowed for achievement of an efficient sorption of palladium ions and uniform distribution of palladium nanoparticles in size and in the pores of the carrier. Catalysts based on PAF-20-G0 and PAF-30-G0 materials demonstrated a high catalytic activity and stability. Nanoparticles in these catalysts were located both on the surface of the support and in the pore space.

Modification of the G1 type led to blockage of the pores of the aromatic framework, which prevented the diffusion of palladium ions into the pores of the carrier. The G1 series catalysts contained larger particles (average size 7–8 nm) and were characterized by a broader size distribution, and most of the nanoparticles were located on the material surface. The activity of the Pd-PAF-20-G1 and Pd-PAF-30-G1 catalysts turned out to be lower than that of the G0 series catalysts, which was due to the blockage of pores.

Thus, to enhance the activity of catalysts, it is necessary to solve the problem of a significant decrease in the porosity of hybrid materials, as well as to improve approaches to the modification of supports. Due to this, it will be possible to significantly expand the field of application of these materials, in particular, to use them to create catalysts for other catalytic processes (processing of petroleum fractions, bio-raw materials, fine organic synthesis, etc.).

**Author Contributions:** E.K., performed conceptualization; M.T., developed methodology; D.M., M.K., performed investigation; Y.K., performed formal analysis; A.M., performed supervision; L.K., performed writing. All authors have read and agreed to the published version of the manuscript.

**Funding:** This research was funded by RSF, grant number 20-19-00380.

**Conflicts of Interest:** The authors declare no conflict of interest.

## References

1. Li, L.; Zhou, C.; Zhao, H.; Wang, R. Spatial control of palladium nanoparticles in flexible click-based porous organic polymers for hydrogenation of olefins and nitrobenzene. *Nano Res.* **2015**, *8*, 709–721. [CrossRef]
2. Li, L.; Zhao, H.; Wang, R. Tailorable synthesis of porous organic polymers decorating ultrafine palladium nanoparticles for hydrogenation of olefins. *ACS Catal.* **2015**, *5*, 948–955. [CrossRef]
3. Garg, G.; Foltran, S.; Favier, I.; Pla, D.; Medina-González, Y.; Gómez, M. Palladium nanoparticles stabilized by novel choline-based ionic liquids in glycerol applied in hydrogenation reactions. *Catal. Today* **2020**, *346*, 69–75. [CrossRef]
4. Chung, J.; Kim, C.; Jeong, H.; Yu, T.; Binh, D.H.; Jang, J.; Lee, J.; Kim, B.M.; Lim, B. Selective semihydrogenation of alkynes on shape-controlled palladium nanocrystals. *Chem. Asian J.* **2013**, *8*, 919–925. [CrossRef] [PubMed]

5. Zhang, Y.; Riduan, S.N. Functional porous organic polymers for heterogeneous catalysis. *Chem. Soc. Rev.* **2012**, *41*, 2083–2094. [CrossRef]
6. Zhang, Q.; Yang, Y.; Zhang, S. Novel functionalized microporous organic networks based on triphenylphosphine. *Chem. Eur. J.* **2013**, *19*, 10024–10029. [CrossRef]
7. Hausoul, P.J.C.; Eggenhuisen, T.M.; Nand, D.; Baldus, M.; Weckhuysen, B.M.; Klein Gebbink, R.J.M.; Bruijnincx, P.C.A. Development of a 4,4′-biphenyl/phosphine-based COF for the heterogeneous Pd-catalysed telomerisation of 1,3-butadiene. *Catal. Sci. Technol.* **2013**, *3*, 2571–2579. [CrossRef]
8. Karakhanov, E.A.; Maksimov, A.L.; Zolotukhina, A.V.; Kardasheva, Y.S. Hydrogenation catalysts based on metal nanoparticles stabilized by organic ligands. *Russ. Chem. Bull.* **2013**, *62*, 1465–1492.
9. Karakhanov, E.; Maximov, A.; Kardasheva, Y.; Semernina, V.; Zolotukhina, A.; Ivanov, A.; Abbott, G.; Rosenberg, E.; Vinokurov, V. Pd nanoparticles in dendrimers immobilized on silica-polyamine composites as catalysts for selective hydrogenation. *ACS Appl. Mater. Interfaces* **2014**, *6*, 8807–8816. [CrossRef]
10. Karakhanov, E.A.; Maximov, A.L.; Zolotukhina, A.V. Selective semi-hydrogenation of phenyl acetylene by Pd nanocatalysts encapsulated into dendrimer networks. *Mol. Catal.* **2019**, *469*, 98–110. [CrossRef]
11. Chen, H.; He, Y.; Pfefferle, L.D.; Pu, W.; Wu, Y.; Qi, S. Phenol Catalytic Hydrogenation over Palladium Nanoparticles Supported on Metal-Organic Frameworks in the Aqueous Phase. *ChemCatChem* **2018**, *10*, 2558–2570. [CrossRef]
12. Jansen, J.F.G.A.; De Brabander-Van Den Berg, E.M.M.; Meijer, E.W. Encapsulation of guest molecules into a dendritic box. *Science* **1994**, *266*, 1226–1229. [CrossRef] [PubMed]
13. Niu, Y.; Crooks, R.M. Dendrimer-encapsulated metal nanoparticles and their applications to catalysis. *C. R. Chim.* **2003**, *6*, 1049–1059. [CrossRef]
14. Yates, C.R.; Hayes, W. Synthesis and applications of hyperbranched polymers. *Eur. Polym. J.* **2004**, *40*, 1257–1281. [CrossRef]
15. Karakhanov, E.A.; Maximov, A.L.; Skorkin, V.A.; Zolotukhina, A.V.; Smerdov, A.S.; Tereshchenko, A.Y. Nanocatalysts based on dendrimers. *Pure Appl. Chem.* **2009**, *81*, 2013–2023. [CrossRef]
16. Karakhanov, E.A.; Maximov, A.L.; Zakharyan, E.M.; Zolotukhina, A.V.; Ivanov, A.O. Palladium nanoparticles on dendrimer-containing supports as catalysts for hydrogenation of unsaturated hydrocarbons. *Mol. Catal.* **2017**, *440*, 107–119. [CrossRef]
17. Karakanov, E.A.; Zolotukhina, A.V.; Ivanov, A.O.; Maximov, A.L. Dendrimer-Encapsulated Pd Nanoparticles, Immobilized in Silica Pores, as Catalysts for Selective Hydrogenation of Unsaturated Compounds. *Chem. Open* **2019**, *8*, 358–381. [CrossRef]
18. Krishnan, G.R.; Sreekumar, K. Synthesis and Characterization of Polystyrene Supported Catalytically Active Poly(amidoamine) Dendrimer-Palladium Nanoparticle Conjugates. *Soft Mater.* **2010**, *8*, 114–129. [CrossRef]
19. Alvarez, J.; Sun, L.; Crooks, R.M. Electroactive composite dendrimer films containing thiophene-terminated poly(amidoamine) dendrimers cross-linked by poly(3-methylthiophene). *Chem. Mater.* **2002**, *14*, 3995–4001. [CrossRef]
20. Murugan, E.; Rangasamy, R. Synthesis, characterization, and heterogeneous catalysis of polymer-supported poly(propyleneimine) dendrimer stabilized gold nanoparticle catalyst. *J. Polym. Sci. Part A Polym. Chem.* **2010**, *48*, 2525–2532. [CrossRef]
21. Li, L.; Zhao, H.; Wang, J.; Wang, R. Facile fabrication of ultrafine palladium nanoparticles with size- and location-control in click-based porous organic polymers. *ACS Nano* **2014**, *8*, 5352–5364.
22. Karakhanov, E.; Kardasheva, Y.; Kulikov, L.; Maximov, A.; Zolotukhina, A.; Vinnikova, M.; Ivanov, A. Sulfide catalysts supported on porous aromatic frameworks for naphthalene hydroprocessing. *Catalysts* **2016**, *6*, 1–11.
23. Yuan, Y.; Zhu, G. Porous Aromatic Frameworks as a Platform for Multifunctional Applications. *ACS Cent. Sci.* **2019**, *5*, 409–418. [PubMed]
24. Maximov, A.; Zolotukhina, A.; Kulikov, L.; Kardasheva, Y.; Karakhanov, E. Ruthenium catalysts based on mesoporous aromatic frameworks for the hydrogenation of arenes. *React. Kinet. Mech. Catal.* **2016**, *117*, 729–743.
25. Wang, F.; Mielby, J.; Richter, F.H.; Wang, G.; Prieto, G.; Kasama, T.; Weidenthaler, C.; Bongard, H.J.; Kegnæs, S.; Fürstner, A.; et al. A polyphenylene support for pd catalysts with exceptional catalytic activity. *Angew. Chem. Int. Ed.* **2014**, *53*, 8645–8648.
26. Kulikov, L.A.; Terenina, M.V.; Kryazheva, I.Y.; Karakhanov, E.A. Unsaturated-compound hydrogenation nanocatalysts based on palladium and platinum particles immobilized in pores of mesoporous aromatic frameworks. *Pet. Chem.* **2017**, *57*, 222–229.

27. Su, J.; Chen, J.S. Synthetic porous materials applied in hydrogenation reactions. *Microporous Mesoporous Mater.* **2017**, *237*, 246–259.
28. Ben, T.; Ren, H.; Shengqian, M.; Cao, D.; Lan, J.; Jing, X.; Wang, W.; Xu, J.; Deng, F.; Simmons, J.M.; et al. Targeted synthesis of a porous aromatic framework with high stability and exceptionally high surface area. *Angew. Chem. Int. Ed.* **2009**, *48*, 9457–9460.
29. Garibay, S.J.; Weston, M.H.; Mondloch, J.E.; Colón, Y.J.; Farha, O.K.; Hupp, J.T.; Nguyen, S.T. Accessing functionalized porous aromatic frameworks (PAFs) through a de novo approach. *CrystEngComm* **2013**, *15*, 1515–1519.
30. Tian, Y.; Song, J.; Zhu, Y.; Zhao, H.; Muhammad, F.; Ma, T.; Chen, M.; Zhu, G. Understanding the desulphurization process in an ionic porous aromatic framework. *Chem. Sci.* **2019**, *10*, 606–613. [PubMed]
31. Vilian, A.T.E.; Puthiaraj, P.; Kwak, C.H.; Hwang, S.K.; Huh, Y.S.; Ahn, W.S.; Han, Y.K. Fabrication of Palladium Nanoparticles on Porous Aromatic Frameworks as a Sensing Platform to Detect Vanillin. *ACS Appl. Mater. Interfaces* **2016**, *8*, 12740–12747. [CrossRef] [PubMed]
32. Yang, Y.; Wang, T.; Jing, X.; Zhu, G. Phosphine-based porous aromatic frameworks for gold nanoparticle immobilization with superior catalytic activities. *J. Mater. Chem. A* **2019**, *7*, 10004–10009. [CrossRef]
33. Nikolaev, S.A.; Zanaveskin, L.N.; Smirnov, V.V.; Averyanov, V.A.; Zanaveskin, K.L. Catalytic hydrogenation of alkyne and alkadiene impurities from alkenes. Practical and theoretical aspects. *Russ. Chem. Rev.* **2009**, *78*, 231–247. [CrossRef]
34. Mallat, T.; Baiker, A. Selectivity enhancement in heterogeneous catalysis induced by reaction modifiers. *Appl. Catal. A Gen.* **2000**, *200*, 3–22. [CrossRef]
35. Xing, R.; Liu, Y.; Wu, H.; Li, X.; He, M.; Wu, P. Preparation of active and robust palladium nanoparticle catalysts stabilized by diamine-functionalized mesoporous polymers. *Chem. Commun.* **2008**, *47*, 6297–6299. [CrossRef]
36. Karakhanov, E.; Maximov, A.; Terenina, M.; Vinokurov, V.; Kulikov, L.; Makeeva, D.; Glotov, A. Selective hydrogenation of terminal alkynes over palladium nanoparticles within the pores of amino-modified porous aromatic frameworks. *Catal. Today* **2019**. [CrossRef]
37. Verde-Sesto, E.; Merino, E.; Rangel-Rangel, E.; Corma, A.; Iglesias, M.; Sánchez, F. Postfunctionalized Porous Polymeric Aromatic Frameworks with an Organocatalyst and a Transition Metal Catalyst for Tandem Condensation-Hydrogenation Reactions. *ACS Sustain. Chem. Eng.* **2016**, *4*, 1078–1084. [CrossRef]
38. Li, L.; Chen, Z.; Zhong, H.; Wang, R. Urea-based porous organic frameworks: Effective supports for catalysis in neat water. *Chem. Eur. J.* **2014**, *20*, 3050–3060. [CrossRef]
39. Zhong, H.; Gong, Y.; Zhang, F.; Li, L.; Wang, R. Click-based porous organic framework containing chelating terdentate units and its application in hydrogenation of olefins. *J. Mater. Chem. A* **2014**, *2*, 7502–7508. [CrossRef]
40. Tang, D.; Sun, X.; Zhao, D.; Zhu, J.; Zhang, W.; Xu, X.; Zhao, Z. Nitrogen-Doped Carbon Xerogels Supporting Palladium Nanoparticles for Selective Hydrogenation Reactions: The Role of Pyridine Nitrogen Species. *ChemCatChem* **2018**, *10*, 1291–1299. [CrossRef]
41. Méry, D.; Astruc, D. Dendritic catalysis: Major concepts and recent progress. *Coord. Chem. Rev.* **2006**, *250*, 1965–1979. [CrossRef]
42. Lu, S.; Hu, Y.; Wan, S.; McCaffrey, R.; Jin, Y.; Gu, H.; Zhang, W. Synthesis of Ultrafine and Highly Dispersed Metal Nanoparticles Confined in a Thioether-Containing Covalent Organic Framework and Their Catalytic Applications. *J. Am. Chem. Soc.* **2017**, *139*, 17082–17088. [CrossRef] [PubMed]
43. Wang, Q.; Tsumori, N.; Kitta, M.; Xu, Q. Fast Dehydrogenation of Formic Acid over Palladium Nanoparticles Immobilized in Nitrogen-Doped Hierarchically Porous Carbon. *ACS Catal.* **2018**, *8*, 12041–12045. [CrossRef]
44. Neeli, C.K.P.; Puthiaraj, P.; Lee, Y.R.; Chung, Y.M.; Baeck, S.H.; Ahn, W.S. Transfer hydrogenation of nitrobenzene to aniline in water using Pd nanoparticles immobilized on amine-functionalized UiO-66. *Catal. Today* **2018**, *303*, 227–234. [CrossRef]
45. Sadjadi, S.; Koohestani, F. Pd immobilized on polymeric network containing imidazolium salt, cyclodextrin and carbon nanotubes: Efficient and recyclable catalyst for the hydrogenation of nitroarenes in aqueous media. *J. Mol. Liq.* **2020**, *301*, 112414. [CrossRef]
46. Zhou, S.; Shang, L.; Zhao, Y.; Shi, R.; Waterhouse, G.I.N.; Huang, Y.; Zheng, L.; Zhang, T. Pd Single-Atom Catalysts on Nitrogen-Doped Graphene for the Highly Selective Photothermal Hydrogenation of Acetylene to Ethylene. *Adv. Mater.* **2019**, *31*, 1900509. [CrossRef]

47. Crooks, R.M.; Zhao, M.; Sun, L.; Chechik, V.; Yeung, L.K. Dendrimer-encapsulated metal nanoparticles: Synthesis, characterization, and applications to catalysis. *Acc. Chem. Res.* **2001**, *34*, 181–190. [CrossRef]
48. Boronoev, M.P.; Zolotukhina, A.V.; Ignatyeva, V.I.; Terenina, M.V.; Maximov, A.L.; Karakhanov, E.A. Palladium Catalysts Based on Mesoporous Organic Materials in Semihydrogenation of Alkynes. *Macromol. Symp.* **2016**, *363*, 57–63. [CrossRef]
49. King, A.S.H.; Twyman, L.J. Heterogeneous and solid supported dendrimer catalysts. *J. Chem. Soc. Perkin* **2002**, *2*, 2209–2218. [CrossRef]
50. Soğukömeroğulları, H.G.; Karataş, Y.; Celebi, M.; Gülcan, M.; Sönmez, M.; Zahmakiran, M. Palladium nanoparticles decorated on amine functionalized graphene nanosheets as excellent nanocatalyst for the hydrogenation of nitrophenols to aminophenol counterparts. *J. Hazard. Mater.* **2019**, *369*, 96–107. [CrossRef]
51. Yang, J.; Yuan, M.; Xu, D.; Zhao, H.; Zhu, Y.; Fan, M.; Zhang, F.; Dong, Z. Highly dispersed ultrafine palladium nanoparticles encapsulated in a triazinyl functionalized porous organic polymer as a highly efficient catalyst for transfer hydrogenation of aldehydes. *J. Mater. Chem. A* **2018**, *6*, 18242–18251. [CrossRef]
52. Xu, D.; Wang, F.; Yu, G.; Zhao, H.; Yang, J.; Yuan, M.; Zhang, X.; Dong, Z. Aminal-based Hypercrosslinked Polymer Modified with Small Palladium Nanoparticles for Efficiently Catalytic Reduction of Nitroarenes. *ChemCatChem* **2018**, *10*, 4569–4577. [CrossRef]
53. Yuan, Y.; Sun, F.; Ren, H.; Jing, X.; Wang, W.; Ma, H.; Zhao, H.; Zhu, G. Targeted synthesis of a porous aromatic framework with a high adsorption capacity for organic molecules. *J. Mater. Chem.* **2011**, *21*, 13498–13502. [CrossRef]
54. Gangadharan, D.; Dhandhala, N.; Dixit, D.; Thakur, R.S.; Popat, K.M.; Anand, P.S. Investigation of solid supported dendrimers for water disinfection. *J. Appl. Polym. Sci.* **2012**, *124*, 1384–1391. [CrossRef]
55. Lu, W.; Sculley, J.P.; Yuan, D.; Krishna, R.; Wei, Z.; Zhou, H.C. Polyamine-tethered porous polymer networks for carbon dioxide capture from flue gas. *Angew. Chem. Int. Ed.* **2012**, *51*, 7480–7484. [CrossRef]
56. Law, R.V.; Sherrington, D.C.; Snape, C.E.; Ando, I.; Korosu, H. Solid State 13C MAS NMR Studies of Anion Exchange Resins and Their Precursors. *Ind. Eng. Chem. Res.* **1995**, *34*, 2740–2749. [CrossRef]
57. Rangel-Rangel, E.; Verde-Sesto, E.; Rasero-Almansa, A.M.; Iglesias, M.; Sánchez, F. Porous aromatic frameworks (PAFs) as efficient supports for N-heterocyclic carbene catalysts. *Catal. Sci. Technol.* **2016**, *6*, 6037–6045. [CrossRef]
58. Thommes, M.; Kaneko, K.; Neimark, A.V.; Olivier, J.P.; Rodriguez-Reinoso, F.; Rouquerol, J.; Sing, K.S.W. Physisorption of gases, with special reference to the evaluation of surface area and pore size distribution (IUPAC Technical Report). *Pure Appl. Chem.* **2015**, *87*, 1051–1069. [CrossRef]
59. Jeromenok, J.; Weber, J. Restricted access: On the nature of adsorption/desorption hysteresis in amorphous, microporous polymeric materials. *Langmuir* **2013**, *29*, 12982–12989. [CrossRef]

© 2020 by the authors. Licensee MDPI, Basel, Switzerland. This article is an open access article distributed under the terms and conditions of the Creative Commons Attribution (CC BY) license (http://creativecommons.org/licenses/by/4.0/).

Article

# Transition Metal Sulfides- and Noble Metal-Based Catalysts for N-Hexadecane Hydroisomerization: A Study of Poisons Tolerance

**Aleksey Pimerzin** [1,2,*], **Aleksander Savinov** [2], **Anna Vutolkina** [1,3], **Anna Makova** [1], **Aleksandr Glotov** [1], **Vladimir Vinokurov** [1] and **Andrey Pimerzin** [2]

[1] Gubkin Russian State University of Oil and Gas, Department of Chemical Technology and Ecology, Division of Physical and Colloid Chemistry, 65 Leninsky Prosp, 119991 Moscow, Russia; annavutolkina@mail.ru (A.V.); ankamackova@yandex.ru (A.M.); glotov.a@gubkin.ru (A.G.); vinok_ac@mail.ru (V.V.)

[2] Samara State Technical University, Department of Chemical Technology, Division of Chemical Technology of Oil and Gas Refinery, 244 Molodogvardeyskaya Street, 443100 Samara, Russia; savinooov13@gmail.com (A.S.); pimerzin.aa@samgtu.ru (A.P.)

[3] Lomonosov Moscow State University, Department of Chemistry, Division of Petroleum Chemistry and Organic Catalysis, 3, 1 Leninskie Gory, 119991 Moscow, Russia

* Correspondence: al.pimerzin@gmail.com; Tel.: +7-846-242-3580

Received: 24 April 2020; Accepted: 21 May 2020; Published: 26 May 2020

**Abstract:** Bifunctional catalysts on the base of transition metal sulfides (CoMoS and NiWS) and platinum as noble metal were synthesized via wetness impregnation of freshly synthesized $Al_2O_3$-SAPO-11 composites, supported with favorable acidic properties. The physical-chemical properties of the prepared materials were studied by X-ray diffraction (XRD), scanning electron microscopy (SEM), low-temperature $N_2$ adsorption and high resolution transmission electron microscopy (HR TEM) methods. Catalytic properties were studied in n-hexadecane isomerization using a fixed-bed flow reactor. The catalytic poisons tolerance of transition metal sulfides (TMS)- and Pt-catalysts has been studied for sulfur and nitrogen, with the amount of 10–100 ppm addition to feedstock. TMS-catalysts show good stability during sulfur-containing feedstock processing, whereas Pt-catalyst loses much of its isomerization activity. Nitrogen-containing compounds in the feedstock has a significant impact on the catalytic activity of both TMS and Pt-based catalysts.

**Keywords:** hydroisomerization; n-alkanes; bifunctional catalysts; stability; CoMoS; NiWS; Pt; catalytic poisons

## 1. Introduction

Nowadays, transformation of n-alkanes to corresponding branched forms by hydroconversion, i.e., hydrocracking or hydroisomerization, is becoming more important for the petroleum refinery and petrochemical industry [1–5]. This process plays a significant role in the production of high-quality gasoline and diesel fuels, as well as motor oils. Diesel and gasoline oil fraction treatment transforms the high freezing long-chain n-paraffins, and contributes to the improvement of the octane number of gasoline and the low-temperature performance of diesel or lubricating oils [6–10]. One of hydroisomerization's advantages is the high yield of liquid products. These light fractions do not have any sulfur, but contain branched alkanes with high octane numbers, which causes isomerization products to be the most valuable component of gasoline. Hydroisomerization of linear long-chain hydrocarbons is becoming more important considering ecological restrictions [11,12].

Hydroisomerization catalysts are bifunctional systems, containing metal particles acting as hydrogenation/dehydrogenation sites and acidic mesoporous support, providing acid sites for isomerization [13,14]. The hydrogenation/dehydrogenation activity of hydroisomerization catalysts is provided by the metallic sites, typically Pt, supported on acidic mesoporous carriers [13,15–17]. Different kinds of zeolites, natural clays and aluminosilicates are widely used as supports for bifunctional catalysts [18–24].

According to widely acknowledged hydroisomerization mechanisms, n-alkanes are dehydrogenated on metallic active sites to form alkene intermediates. The resulted olefins are protonated and isomerized mainly on Brønsted acid sites, followed by rapid hydrogenation over metal sites to the corresponding branched iso-alkanes (Figure 1) [25,26]. Besides, the length of the hydrocarbon chain is considered to be correlated with the reactivity thereof [27]. Thus, the acidity properties are inversely proportional to the chain length [1]. Therefore, the selectivity of the catalyst towards isomerization is affected by a proper balance between metallic and acidic sites.

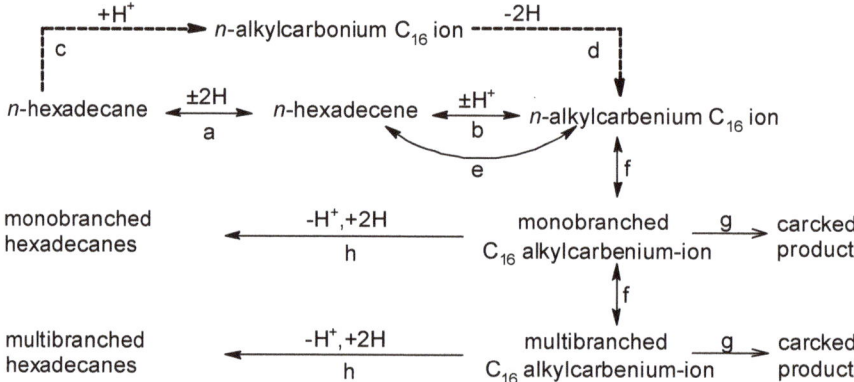

**Figure 1.** Scheme of n-hexadecane hydroisomerization. **a**—Hydrogenation-dehydrogenation on metallic sites; **b**—Protonation-deprotonation on acid sites; **c**—Addition of proton to form alkylcarbonium ion on acid; **d**—Dehydrogenation to form alkylcarbenium ion; **e**—Competitive adsorption–desorption of alkene and carbenium ion on acid sites; **f**—Rearrangement of alkylcarbenium ion; **g**—Cracking of alkylcarbenium ion; **h**—Hydrogenation to form alkane on metallic sites.

The structural properties of the support also affect the shape-selectivity of molecular sieves. In general, the smaller the zeolite pore size, the lower the methyl branching [28,29]. While for microporous molecular sieves methyl branching takes place, for wide pore openings and large cavities, ethyl and propyl branched isomers are obtained [7,30]. Multibranched isomers formed during n-paraffins isomerization are more susceptible to hydrocracking, which leads thereby to lower isomerization selectivity [31]. The support structure and textural properties are also crucial for long-chain isomerization from the point of view of diffusion limitations, due to the transformation of linear alkanes occurring in the micropores: the rapid diffusion of molecules to the bulk phase should be provided before the undesired cracking reactions occur.

The hydrogenation/dehydrogenation activity of hydroisomerization catalysts is provided by metallic sites, typically Pt, supported on acidic mesoporous carriers [13,15]. Different kinds of zeolites, natural clays and aluminosilicates are widely used as supports for bifunctional catalysts [18–24]. The most significant characteristic for these catalysts from an industrial viewpoint is stability to poisons [32,33]. Noble metal-based catalysts stand out as more active for hydroisomerization when the content of the active metals is higher than 0.5 wt %, which makes them very expensive [34–37]. Moreover, catalyst poisoning by nitrogen- and sulfur-containing compounds leads to dramatic decreases in activity [38–41]. Metal active sites poisoning by sulfur involves hydrogen sulfide adsorption followed by its dissociation, leading to rearrangement of the metal surface, wherein the polysulfide layer forms, and catalyst deactivation is reversible. When the S/Pt ratio is higher than 0.4, the poisoning becomes irreversible. This may also be caused by strong Pt–S covalent bonding [42]. The deactivation of the noble metal-based catalysts by nitrogen-containing compounds takes place due to their strong adsorption on the acid sites of the support. As a result, the content of available active sites decreases, which has a negative influence on catalytic activity [43]. Therefore, the industrial application of noble metal-based catalysts for hydroisomerization is limited, and research focuses shift towards novel low-cost bifunctional catalysts with sufficient activity and stability. As such, the development of transition metal-containing catalysts for hydroisomerization of linear alkanes is of both scientific and industrial interest [44–47].

The most promising sulfur tolerance catalysts are transition metal sulfides. The transition metal sulfides providing high catalytic activity are widely used for hydrotreatment processes [44,48–51]. CoMo bimetallic catalysts have been investigated for ethylbenzene hydrogenation [34]. Appropriate acidity of the support favors the hydrogenation/isomerization activity of the catalysts, which allows the performance of ultra-deep sulfur removal from oil fractions [52–54].

A number of microporous zeolites, such as ZSM-22, ZSM-23 and ZSM-48, were used as functional supports for hydroisomerization catalysts. Among one-dimensional porous materials, structured SAPO-11 exhibits appropriate acidity and shape-selectivity, making it one of the more promising supports for hydroisomerization catalysts. This silicoaluminophosphate, consisting of $Al_2O_3$, $SiO_2$ and $P_2O_5$ as the main components, has a one-dimensional, 10-membered ring channel system, with an elliptical pore opening of 0.44 × 0.64 nm [55]. Thanks to moderate acidity and shape-selectivity, the medium-pore SAPO-11 acts as a functional support for isomerization catalysts, favoring selective isomerization of long-chain alkanes (for monomethyl isomers) [56]. Thus, SAPO-11-supported Pt-catalysts were found to exhibit high selectivity to skeletal isomers in n-dodecane hydroisomerization at 300 °C and hydrogen pressure of 1 bar (weight hourly space velocity (WHSV) = 2 hr$^{-1}$) [57]. Similar results were obtained in [58], when 1 wt % of Pt was loaded on SAPO-11 and tested at 307 °C under higher hydrogen pressure (20 bar). Ni-Mo-SAPO-11-catalysts have demonstrated a high selectivity to $C_{18}$ hydrocarbons (57%) at 350 °C under 30 bar [45]. More recently, the SAPO-11-supported CoMo-catalyst has been applied in n-hexadecane hydroisomerization, and exhibited comparable activity with noble metals-based catalysts [59].

Herein, we examined the stability of transition metal-containing catalysts in terms of the isomerization function of n-alkanes to poisoning by nitrogen and sulfur compounds. This investigation also includes calculations of kinetic parameters, evaluating the n-paraffin isomerization selectivity depending on sulfur and nitrogen influence.

## 2. Results

The key factor, from the industrial viewpoint, in catalyst application, is mechanical stability, provided by supports. In general, hydroisomerization catalysts are formed into pellets by the extrusion of functional acidic components mixed with boehmite. The latter provides $\gamma$-$Al_2O_3$ formation after calcination at 550–600 °C [15,59]. Thus, we here report the examination of pelletized $Al_2O_3$-SAPO-11-supported Pt and transition metal sulfide catalysts in the hydroisomerization of n-hexadecane in the presence of poisons. The SAPO-11 component in these catalysts has appropriate acidity and provides selectivity in hydroisomerization reactions.

*2.1. Physical-Chemical Properties of the Solids*

Figure S1 shows the diffraction pattern of synthesized $Al_2O_3$-SAPO-11 support. Observed signals on the recorded XRD pattern for $Al_2O_3$-SAPO-11 material are consistent with standard SAPO-11 (PDF 047-0614) and $Al_2O_3$ (PDF 075-0921).

Figure S2 shows the SEM image of the prepared $Al_2O_3$-SAPO-11 support. SEM images of the composite support surface looks like a mixture of alumina, with a smooth surface and well-dispersed crystals of silicoaluminophosphate, as can be seen from energy dispersive X-rays analysis (EDX) mapping (Figure 2). SAPO-11 particles (1–5 μm) are uniformly distributed over the $Al_2O_3$ surface. The structure of the SAPO-11 crystals does not look damaged by acid.

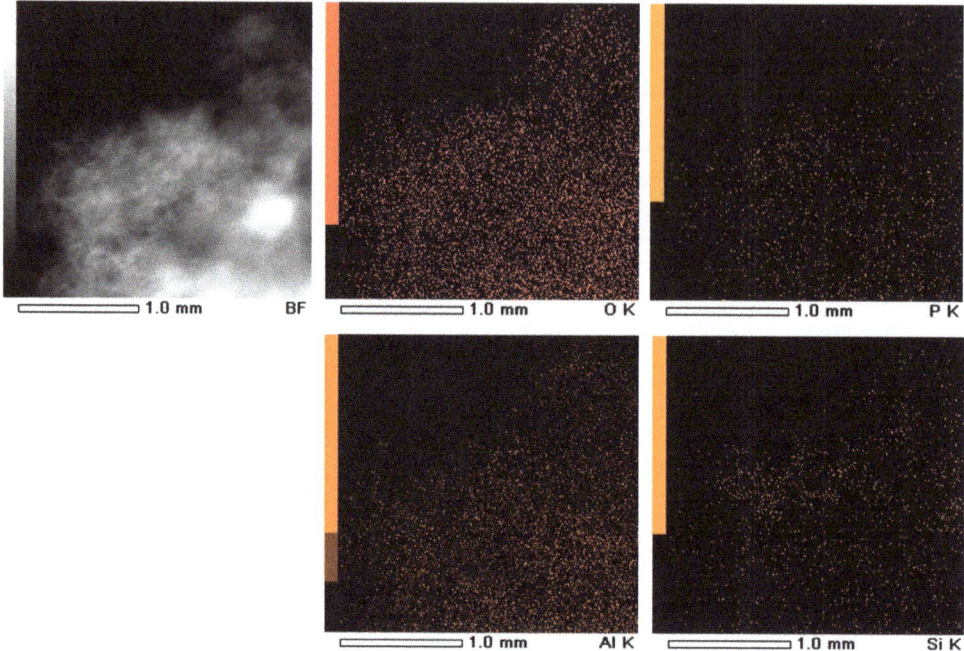

**Figure 2.** EDX image mapping of the elements on composite $Al_2O_3$-SAPO-11 support.

The textural properties of the prepared $Al_2O_3$-SAPO-11 composite support and bifunctional catalysts were measured by low-temperature $N_2$ adsorption. The nitrogen adsorption–desorption isotherms and pore size distributions of the support and synthesized catalysts are shown in Figure 3. Synthesized materials exhibited a typical type I–type IV combination of isotherms, indicating a micro-mesoporous starting material. The total nitrogen adsorbed is slightly higher for the support, according to its higher pore volume. In particular, the steep uptake at lower relative pressure ($P/P_0 < 0.05$) indicates the $N_2$ filling in the micropores of the SAPO-11 phase. Meanwhile, all the prepared catalysts and composite supports show isotherms with an obvious H3 + H4 hysteresis loop, which is the result of the composite support's nature, combining the micropores of SAPO-11 and the slit-shaped pores of alumina.

The textural properties of the prepared $Al_2O_3$-SAPO-11 composite supports and transition metal sulfide (TMS) catalysts, as well as the reference Pt-catalyst, are listed in Table 1. All the prepared materials offer bimodal characteristics of pore size distribution (Figure 3). Pores with an average diameter about 3.8 nm (t-plot) correspond to SAPO-11, and the other ones, with 8.0 nm diameter, are related to alumina. Active metals loading has no significant effect on the textural properties of the supported catalysts. Specific surface area decreases from 268 to about 200 $m^2/g$ in the case of sulfide catalysts synthesis, and keeps almost constant for Pt-catalyst. The average pore volume and micropore properties evaluated, by t-plot and density functional theory (DFT) methods, do not change after catalysts' impregnation (Table 1).

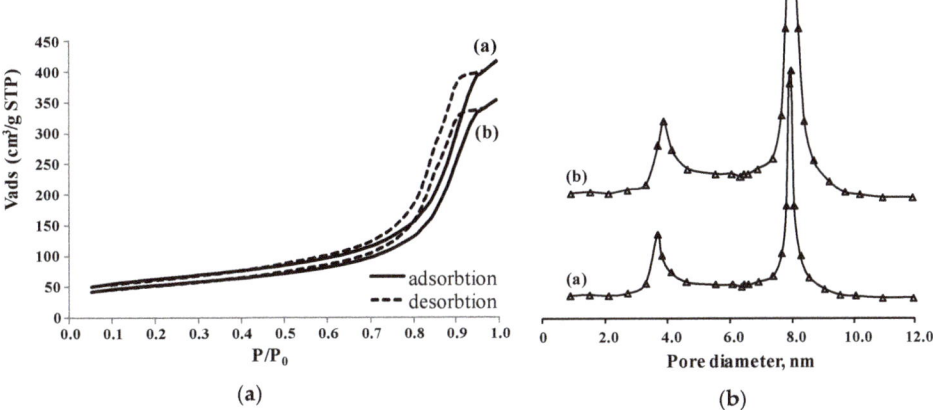

**Figure 3.** N$_2$ adsorption isotherms (a) and pore size distribution curves (b) of composite support (curves a) and prepared catalysts (curves b).

**Table 1.** Textural properties of the prepared solids.

| Catalyst (Support) | Brunauer–Emmett–Teller (BET) | | | t-Plot & DFT | | |
|---|---|---|---|---|---|---|
| | S$_{BET}$ (m$^2$/g) | D$_p$ avr. (nm) | V$_{pore}$ (cm$^3$/g) | S$_{micro}$ (m$^2$/g) | D$_{micro}$ (nm) | V$_{micro}$ (cm$^3$/g) |
| Al$_2$O$_3$-SAPO-11 (support) | 268 | <4.0 & 8.1 * | 0.50 | 69 | 3.8 | 0.06 |
| CoMoS/Al$_2$O$_3$-SAPO-11 | 204 | <4.0 & 8.0 * | 0.49 | 66 | 3.7 | 0.05 |
| NiWS/Al$_2$O$_3$-SAPO-11 | 198 | <4.0 & 7.9 * | 0.47 | 64 | 3.8 | 0.05 |
| Pt/Al$_2$O$_3$-SAPO-11 | 259 | <4.0 & 8.1 * | 0.51 | 59 | 3.7 | 0.04 |

* bimodal pore size distribution.

Figure 4 shows representative HR TEM micrographs of the prepared catalysts. The active phase of CoMoS- and NiWS-catalysts is preceded by the typically well-dispersed sulfide phase—Hexagonal crystallites of Mo(W)S$_2$ decorated with Co(Ni) atoms on the edges, as presented in Topsøe's works [44,60,61]. On the HR TEM pictures, the sulfide phase looks like single or stacked dark stripes. The average length of CoMoS sulfide's active phase is 3.6 nm, and a stacking number of 2.1 resulted in dispersion equal to 0.33 (Table 2). NiWS's active phase is characterized by average length of 4.6 nm and stacking number 1.9, resulting in 0.27 dispersion. The active phase of the reference Pt-catalyst is presented by nanosized metal clusters dispersed on the catalyst surface. The average particle size of Pt clusters on the surface of Pt/Al$_2$O$_3$-SAPO-11-catalyst is 1.5 nm.

The acidic properties of CoMoS and NiWS-based bifunctional catalysts have been studied with the NH$_3$-TPD method, and the amounts of desorbed ammonia are presented in Table 2. The measured acidic properties of sulfide catalysts provide very close values of desorbed NH$_3$ for each type of acid site, and are in accordance with previously obtained results [62]. Sulfide active phase formation leads to an increase in the amount of high-temperature active sites, in comparison with Al$_2$O$_3$-SAPO-11 support, from 0.22 up to 0.62 mmol/g. The NiWS-catalyst exhibits slightly higher total acidity evaluated by NH$_3$-TPD, about 10% relative to CoMoS-catalyst. Catalytic properties of the Pt/Al$_2$O$_3$-SAPO-11-catalyst are almost equal to the initial composite support, most likely due to low concentration of platinum.

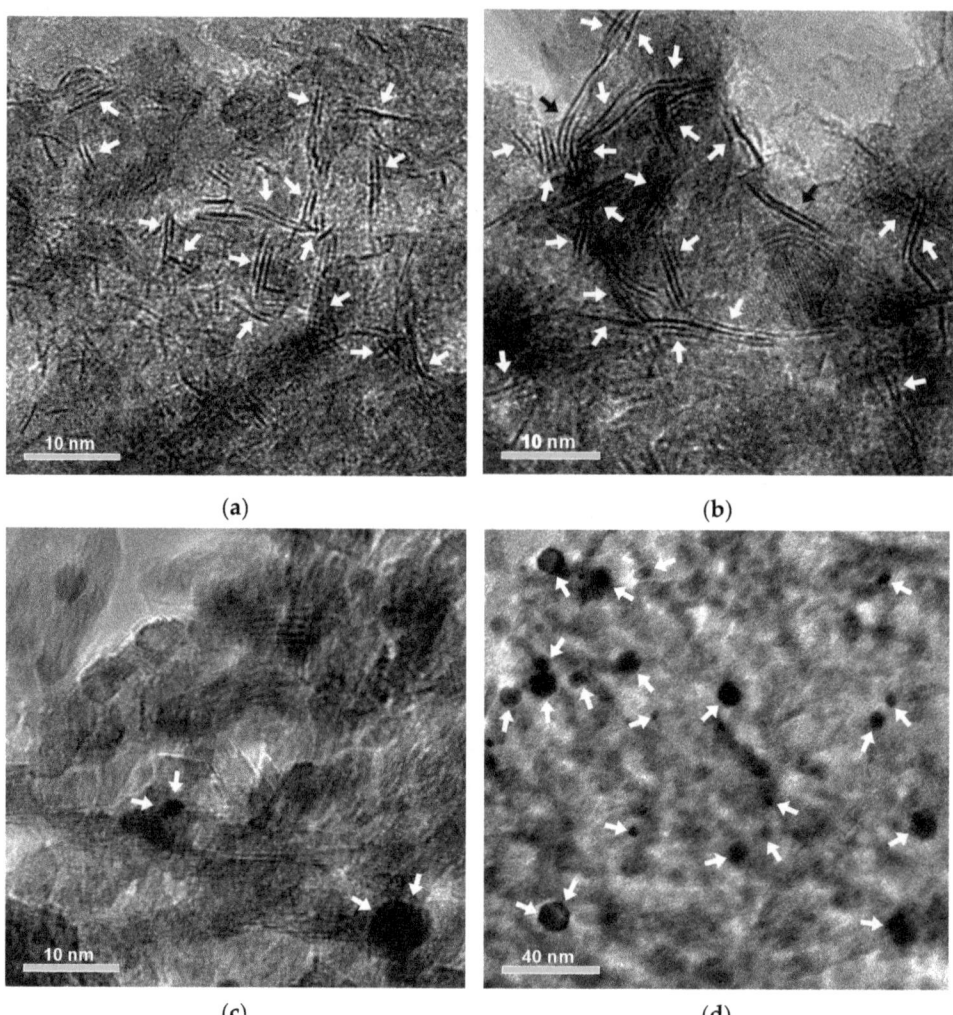

**Figure 4.** HR TEM micrographs of the synthesized bifunctional catalysts: (**a**) CoMoS/Al$_2$O$_3$-SAPO-11, (**b**) NiWS/Al$_2$O$_3$-SAPO-11, (**c**,**d**) Pt/Al$_2$O$_3$-SAPO-11.

Table 2. Metals content and active phase properties of the prepared catalysts.

| Parameter | Catalyst | | |
|---|---|---|---|
| | CoMoS/ Al$_2$O$_3$-SAPO-11 | NiWS/ Al$_2$O$_3$-SAPO-11 | Pt/ Al$_2$O$_3$-SAPO-11 |
| **Active metals loading (wt %)** | | | |
| • Mo(W) or Pt (for ref. catalyst) | 10.6 | 18.5 | 1.0 |
| • Co(Ni) | 3.1 | 3.0 | - |
| **Active phase morphology** | | | |
| • Average length, $\bar{L}$ (nm) (diameter for ref. catalyst) | 3.6 | 4.6 | 1.5 |
| • Average stacking number, $\bar{N}$ | 2.1 | 1.9 | - |
| • Dispersion of active phase, $D$ | 0.33 | 0.27 | 0.91 |
| **Acidity—TPD NH$_3$, mmol/g \*** | | | |
| • Weak | 0.421 | 0.423 | 0.367 |
| • Medium | 0.193 | 0.204 | 0.261 |
| • Strong | 0.507 | 0.616 | 0.236 |
| • Total | 1.121 | 1.243 | 0.865 |

\* Temperature programmed desorption of ammonia (TPD NH$_3$), acid sites are classified as weak (100–250 °C), medium (250–400 °C), and strong (> 400 °C).

## 2.2. Catalytic Properties Examination

It is well-known that Pt-containing bifunctional catalysts are much more active than other ones, nevertheless, several previous studies have shown the possibility of using the transition metal-based catalysts application for n-alkanes hydroisomerization. Figure 5 shows the comparison between the reaction rate constant of n-hexadecane isomerization at various temperatures with CoMoS-, NiWS- and Pt-catalysts, supported with Al$_2$O$_3$-SAPO-11 support. The obtained results correlate with published data [59,62]. The difference in activity between Pt- and TMS-based catalysts gradually decreases as the temperature increases from 300 to 340 °C. The Pt-catalyst demonstrates a reaction rate constant 10 times higher at 300 °C and 4.5 times higher at 340 °C, compared to the NiWS-catalyst (Figure 5). The increased temperature of the process is likely favorable for TMS-catalysts application. Methyl- and ethylpentadecanes were the main products of the isomerization reaction. Coke formation, as well as cracking products, were not observed under the studied conditions for any experiments.

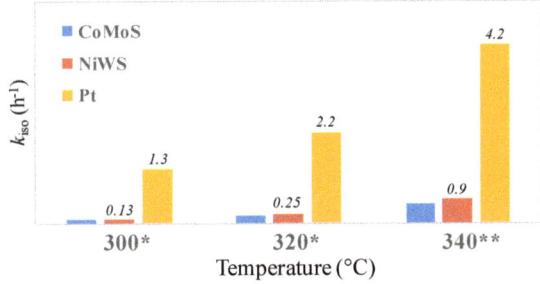

**Figure 5.** Reaction rate constants of n-hexadecane isomerization at 300–340 °C, versus Al$_2$O$_3$-SAPO-11 supported CoMoS-, NiWS- and Pt-catalysts. \*—Results adopted from [59,62], \*\*—Results obtained in this work.

Catalytic stability tests for all catalysts have been performed with sulfur- and nitrogen-containing feeds at constant reaction conditions. Table 3 summarizes the obtained results of catalytic activity measurements of TMS- and Pt-based catalysts. CoMoS- and NiWS-catalysts demonstrate perfect stability during the experiments with sulfur-containing feedstock. The conversion of n-hexadecane keeps constant, 53% and 59% for CoMoS- and NiWS-catalysts, respectively, while processing feeds with up to 100 ppm of sulfur. Reaction rate constants are comparable for both sulfide catalysts in the experiments with sulfur-containing feedstocks, and equal to the initial values for pure n-$C_{16}$ processing. Consequently, the inhibiting effect of sulfur on TMS-catalysts is equal to zero. The Pt-based catalyst is exposed to sulfur poisoning more intensively. The addition of 10 ppm sulfur into the feedstock reduces n-$C_{16}$ conversion from 79% to 52%, which is valued at 53% activity inhibition, but the residual activity nevertheless is higher than for the NiWS- or CoMoS-catalysts. The sulfur concentration increasing up to 100 ppm has a dramatic (77%) inhibiting effect on the Pt-catalyst. The isomerization reaction rate constant is $0.35 \pm 0.006 \times 10^{-4}$ (mol g$^{-1}$ h$^{-1}$), which is almost three times lower than NiWS-catalyst's one. For all experiments, hydroisomerization selectivity was almost 100%, and no cracking products were observed.

**Table 3.** Catalytic performance of TMS-based and reference Pt-catalyst in n-hexadecane hydroisomerization at 340 °C, 1.5 MPa, 150 nL/L $H_2$ to feedstock ratio.

| Catalyst & Reaction Parameter | Reaction Conditions (T = 340 °C, P = 15 bar, $H_2$/Feedstock = 150nL/L) | | | | |
|---|---|---|---|---|---|
| | WHSV (h$^{-1}$) | Feedstock: N-Hexadecane 2.9 wt % + Catalytic Poison | | | |
| | | No Poison | 10 ppmS | 100 ppmS | 10 ppmN | 50 ppmN |
| **CoMoS/Al$_2$O$_3$-SAPO-11** | | | | | | |
| • n-$C_{16}$ conversion (%) | 1.0 | 52.8 | 53.1 | 52.0 | 29.8 | 12.4 |
| • $k_{iso} \times 10^4$ (mol g$^{-1}$ h$^{-1}$) | | 0.76 ± 0.02 | 0.77 ± 0.02 | 0.74 ± 0.01 | 0.36 ± 0.01 | 0.13 ± 0.003 |
| • Inhib. effect | | - | - | 2% | 52% | 83% |
| **NiWS/Al$_2$O$_3$-SAPO-11** | | | | | | |
| • n-$C_{16}$ conversion (%) | 1.0 | 59.0 | 59.5 | 58.8 | 38.8 | 18.8 |
| • $k_{iso} \times 10^4$ (mol g$^{-1}$ h$^{-1}$) | | 0.90 ± 0.01 | 0.92 ± 0.02 | 0.90 ± 0.02 | 0.50 ± 0.01 | 0.21 ± 0.005 |
| • Inhib. effect | | - | - | 1% | 45% | 76% |
| **Pt/Al$_2$O$_3$-SAPO-11** | | | | | | |
| • n-$C_{16}$ conversion (%) | 3.0 | 78.6 | 51.8 | 8.5 | 9.1 | 2.9 |
| • $k_{iso} \times 10^4$ (mol g$^{-1}$ h$^{-1}$) | | 4.75 ± 0.22 | 2.2 ± 0.044 | 0.35 ± 0.006 | 0.29 ± 0.01 | 0.09 ± 0.004 |
| • Inhib. effect | | - | 53% | 77% | 94% | 98% |

Figure 6 is used to show, in a graphic outline, the comparison of residual catalytic activity, i.e., the measured activity after the n-hexadecane conversion stopped decreasing, of bifunctional catalysts during sulfur- and nitrogen-containing feedstock processing. Similar reaction conditions can be used in this case due to the higher inhibition effect of catalytic poisons on Pt-containing catalyst. A reaction rate constant more than two times lower was measured for Pt-catalyst in the experiment with 100 ppm sulfur, resulting in residual n-$C_{16}$ conversion becoming equal to 37%, versus 59% for the NiWS-catalyst, meanwhile, the 10 ppm of sulfur in the feedstock allows the Pt-catalyst to provide higher catalytic activity in n-$C_{16}$ isomerization (Figure 6a).

The calculated inhibition effect of 10 ppm nitrogen-containing feed on CoMoS- and NiWS-catalysts is about 45–52%, meaning that the reaction rate constant is only half of the initial one and the conversion of n-hexadecane decreases to 30% and 39% for CoMoS and NiWS samples, respectively (Table 3). The Pt-catalyst already demonstrates extremely low n-$C_{16}$ conversion at studied reaction conditions when 10 ppm of nitrogen is added. The inhibiting effect is 94% in this case. The increasing of the nitrogen content to 50 ppm in the feedstock dramatically drops n-hexadecane conversion measured for all the prepared catalysts. CoMoS- and NiWS-catalysts lose 83% and 76% of their initial activity, respectively. The observed n-hexadecane conversion over the Pt-catalyst is almost equal to zero; only about 3% of $C_{16}$ isomers are detected in the product. Such low values of n-hexadecane conversion measured for Pt-catalyst in the nitrogen-containing feedstock processing is connected with the high WHSV 3.0 $h^{-1}$, which was chosen for the Pt-catalyst due to the extremely high catalytic activity in pure n-$C_{16}$ hydroisomerization. However, by paying attention to the isomerization reaction rate constants calculated for all the catalysts (Table 3), it becomes possible to conclude that the catalytic behavior of the studied TMS- and Pt-based bifunctional catalysts is quite similar in nitrogen-containing feedstock processing.

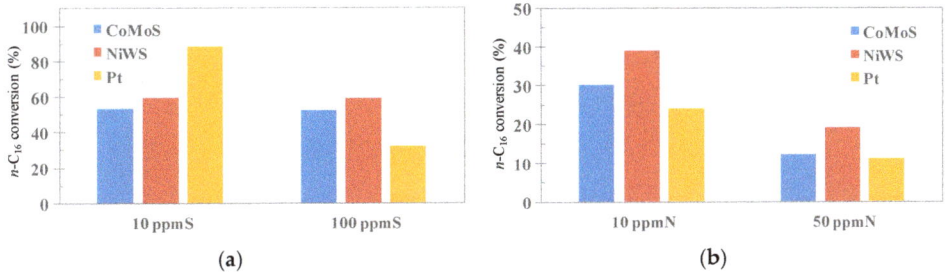

**Figure 6.** Conversion of n-$C_{16}$ (with sulfur (**a**) and nitrogen (**b**) poisoning) over TMS- and Pt-catalysts at similar conditions: (T = 340 °C, P = 15 bar, WHSV = 1.0 $h^{-1}$, $H_2$/feedstock = 150 nL/L).

The comparison of the n-$C_{16}$ conversion with the prepared catalysts under the similar reaction conditions (Figure 6b) demonstrates quite clearly that the isomerization activity of synthesized TMS and- Pt-based bifunctional catalysts was inhibited by nitrogen through the similar mechanism, which differs from the one for Pt-catalysts poisoning by sulfur. The residual catalytic activity is the result of a strong limitation of the isomerization reaction in both TMS- and Pt-based catalysts caused by nitrogen in the feedstock. The logical explanation of such catalytic behavior is that nitrogen mostly poisons acid sites of bifunctional catalysts, leading thereby to decreases of acidity and isomerization activity, while the hydrogen sulphide changes the Pt into sulfide form, decreasing thereby its hydrogenation–dehydrogenation activity [7,33]. Since the TMS-based catalysts have already sulfided the active CoMoS or NiWS phase, sulfur in the feedstock has no inhibiting effect. The acid sites deactivating in the bifunctional catalysts is equally undesirable for both TMS- and Pt-containing catalysts.

In addition to the established difference in the inhibition effect of sulfur and nitrogen on Pt-catalyst activity, the rate of catalyst poisoning by the corresponding feedstock's undesired component was studied (Figure 7). The period of catalytic activity stabilization, in hours, required for reaching the steady state of n-hexadecane conversion was different when sulfur and nitrogen components were introduced into the feedstock. It took about 38 h during sulfur-containing feedstock processing to get the stable conversion of n-hexadecane (region "c" on Figure 7a). The inhibiting effect of nitrogen was exhibited not only in a significant degree, but also more rapidly. It takes about 20 h for all the studied catalysts to get almost-steady n-$C_{16}$ conversion—Region "c", specified in Figure 7b. It is important to point out that catalytic activity stabilization for both TMS- and Pt-based catalysts takes the same period. Almost the same time interval is required to get a steady state when 10 and 50 ppm of nitrogen are introduced into the feedstock.

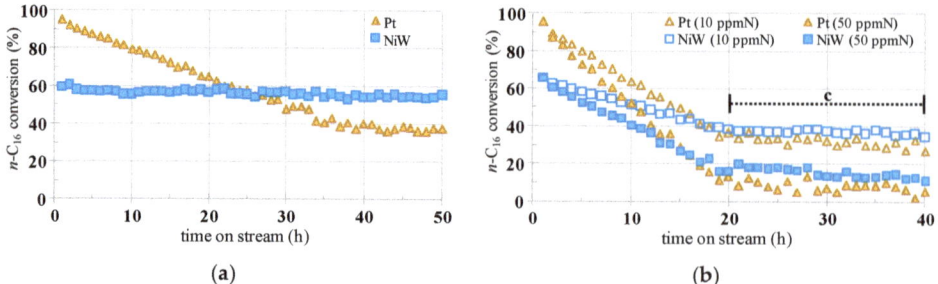

**Figure 7.** The dependence of the n-hexadecane conversion in time for sulfur-containing (**a**) and nitrogen-containing (**b**) feedstock processing.

The observed differences in poisoning kinetics are most likely connected with mechanisms focused on metal active sites in the case of sulfur poisoning of Pt-catalyst, and the acid sites of the silicoaluminophosphate component of the composite supports. In view of this, the NiWS-catalyst provides slightly higher tolerance to nitrogen in feedstock in comparison with CoMoS- and Pt-catalysts. NiWS sulfide's active phase exhibits moderate acidic properties, and is probably involved in the isomerization step of the n-$C_{16}$ reaction.

## 3. Materials and Methods

### 3.1. Preparation of the Solids

SAPO-11 was hydrothermally synthesized according to the reported procedure [63] from the pseudoboehmite (Sasol Company, Hamburg, Germany), orthophosphoric acid (Merck) and Ludox AS30 colloid silica (Aldrich), used as alumina, phosphorus and silicon sources, respectively. The final molar composition of the prepared gel was: 1.0 $Al_2O_3$; 1.0 $P_2O_5$; 0.3 $SiO_2$; 1.2 DPA; 45 $H_2O$. The crystallization was performed in a Teflon-lined stainless steel autoclave at 200 °C for 24 h. The product was water-washed, dried and calcined at 600 °C for 6 h to obtain the SAPO-11 material.

Composite $Al_2O_3$-SAPO-11 support was prepared by mixing SAPO-11 with AlOOH boehmite followed by incipient wetness impregnation with peptizer (concentrated nitric acid with modulus 0.04) and further molding by extrusion to form pellets with 2 mm diameter [15,59]. Pellets were dried at 80 and 110 °C for 4 h at each temperature followed by calcination in air at 550 °C for 6 h.

The bifunctional catalysts were prepared by incipient wetness impregnation technique. Composite $Al_2O_3$-SAPO-11 support was crushed and sieved to particles between 0.25 and 0.50 mm. CoMoS-catalyst was synthesized using 12-molybdophosphoric heteropolyacid (HPA) $H_3[PMo_{12}O_{40}] \cdot nH_2O$ and cobalt (II) carbonate ($CoCO_3$). NiWS-catalyst was prepared from phosphotungstic acid $H_3[PW_{12}O_{40}] \cdot nH_2O$ and nickel (II) carbonate ($NiCO_3$). Citric acid was used as a chelating agent. An $H_2PtCl_6$ solution was used to prepare Pt-based catalysts. Aqueous impregnation solutions were prepared by dissolving all listed components in deionized water under continuous stirring at 50 °C. The amount of dissolving precursors and corresponding concentrations of active metals in solutions were calculated from the required metal loadings (Table 2). For example, for CoMoS/$Al_2O_3$-SAPO-11-catalyst preparation in amount of 50 g, 10.85 g of 12-molybdophosphoric HPA and 3.21 g of cobalt carbonate were dissolved in water with 6.02 g of citric acid. Then 40 g of freshly synthesized $Al_2O_3$-SAPO-11 carrier was impregnated with 36 $cm^3$ of prepared solution. All impregnated materials were dried at 120 °C for 4 h. Pt-catalyst was further calcined at 450 °C (2 h) and reduced just before catalytic testing in a hydrogen atmosphere at 400 °C. TMS-catalysts were activated by sulfidating with a mixture of 15% $H_2S$ in $H_2$ at 400 °C, 1.0 MPa for 2 h.

The elemental analysis (Mo, W, Co, Ni and Pt) of the prepared catalysts was performed by Shimadzu EDX800HS analyzer. SEM-EDX analysis was used to confirm the uniform distribution and check the possible agglomeration of active metals on the surface of the catalysts.

*3.2. Characterization of Supports and Catalysts*

X-ray powder diffraction measurements were performed on a Bruker D2 X-ray Diffraction system using Cu Kα radiation ($\lambda$ = 1.54056 Å) with a scanning speed of 2.0°/min. The powders were loaded on a glass disk by packing the samples into a cavity on the disk. The diffractograms were analyzed using standard JCPDS files.

The surface morphology of the supports was examined by scanning electronic microscopy (SEM) on a Carl Zeiss EVO 50 microscope.

The textural characteristics of the prepared supports and catalysts were measured by low-temperature $N_2$ adsorption at −196 °C on a Quantachrome Autosorb-1 porosimeter after outgassing procedure under deep vacuum at 350 °C for 4 h. Specific surface area (SSA) was calculated using the Brunauer–Emmett–Teller BET method. The total pore volume and pore size distribution were calculated from the desorption branch of the isotherm using the Barrett-Joiner-Halenda (BJH) model. Properties of micropores were evaluated using the *t*-plot method.

HR TEM images of the catalysts were obtained on a JEOL JEM-2100 microscope with electron tube voltage of 200 kV. To measure the key morphological characteristics at least 500 particles from 10–12 regions of the surface were handled for every sample. The average length of Mo(W)S slabs, as well as average number of slabs per stack, were calculated for TMS-catalysts according to the commonly accepted approach. The average diameter of active phase particles was evaluated in the case of Pt-catalyst. The dispersion (D) of active phase was calculated on the basis of geometric characteristics and evaluated as available atoms on the surface of metal clusters, with assumptions that active particles of metallic catalysts are spherical and Mo(W)$S_2$ slabs are perfect hexagons [64–66].

*3.3. Catalytic Activity Examination*

The catalytic tests were performed using a fixed-bed flow reactor. A 2.5 $cm^3$ volume of the catalyst was charged into the isothermal region of the reactor as particles of 0.25–0.50 mm. All catalysts were activated before testing directly in the reactor. Pt-based catalyst was reduced in 0.5 MPa hydrogen atmosphere at 400 °C for 2 h, while TMS samples were sulfided at 400 °C in an atmosphere of $H_2S/H_2$ for 2 h. A mixture of n-hexadecane (2.9 wt %) in n-heptane as a solvent with optional addition of dimethyl disulfide (DMDS) (as sulfur source) and quinolone (as nitrogen source) in 10–100 ppm concentration was introduced into the reactor as a model feedstock for evaluation of hydroisomerization performance and inhibiting effect. N-heptane was not reacted under the experimental conditions.

Catalytic activity was examined under the following conditions: temperature 340 °C, 1.5 MPa $H_2$, 1–3 $h^{-1}$ (weight hourly space velocity (WHSV), and a 150 nL/L volume ratio of $H_2$ to feedstock. The liquid product compositions of the samples collected every 1.0 h were determined using an Agilent 7890A Gas Chromatograph equipped with a Flame Ionization Detector and 30 m DB-5 Fused Silica capillary column (30 m * 0.32 mm * 0.5 µm). Volume of injection was equal to 1.0 µm, the temperature of the injector and detector was 250 °C. The program of temperature started from 60 °C and was raised to 180 °C with 5 °C/min rate. Helium was used as carrier-gas. An example chromatogram is presented in Figure S3. The identification of the obtained reaction products was performed using gas chromatography/mass-spectrometry analysis and by matching retention times with available commercial standards. Methyl- and ethylpentadecanes were the main products of the reaction. The corresponding peaks on the chromatograms were located in the interval of 7.6–8.4 min retention time.

The hydroisomerization reaction was allowed to proceed for the period of time that provided the steady state of the process. In the case of sulfur and nitrogen addition into the feedstock this period varied from 20 to 38 h.

Conversion and isomerization selectivity were calculated using the following equations:

$$x_{iso} = \frac{\text{consumed } n-\text{hexadecane}}{\text{initial } n-\text{hexadecane content}} = \frac{\text{iso} - \text{hexadecanes}}{\text{initial } n-\text{hexadecane content}}, \quad (1)$$

The reaction rate constant of the hydroisomerization was determined using the following equation:

$$k_{iso} = -\frac{F}{m}\ln(1-x_{iso}), \quad (2)$$

where $k_{iso}$ is the pseudo-first-order reaction constant for the n-hexadecane hydroisomerization (mol g$^{-1}$ h$^{-1}$); $x_{iso}$ is the conversion (%) of n-hexadecane into the C$_{16}$ isomers; $F$ is the reactant (n-hexadecane) flow (mol h$^{-1}$) and $m$ is the weight of the catalyst (g).

The inhibiting effect was defined as the ratio of the reaction rate constant after reaching the steady state with poison-containing feedstock processing and the initial reaction rate constant.

$$E_{inh} = \frac{k_{iso}}{k_{iso}^0}, \quad (3)$$

where $k_{iso}^0$ is the initial reaction rate constant during pure n-hexadecane isomerization and $k_{iso}$ is the reaction rate constant for poison-containing feedstock processing.

The residual catalytic activity declared in the manuscript refers to the values of catalytic activity measured after the n-hexadecane conversion stopped decreasing, in the cases of sulfur- or nitrogen-containing feedstock.

## 4. Conclusions

Transition metal sulfide (CoMoS and NiWS) and platinum-containing bifunctional catalysts were prepared by a incipient wetness impregnation technique, using a freshly synthesized Al$_2$O$_3$-SAPO-11 composite support with favorable acidic properties. The structure of the support was confirmed by powder XRD and SEM techniques. All the solids have comparable specific surface areas, from 200 to about 270 m$^2$/g, and the bimodal pore size distributions, with maximums at 3.8 and 8.0 nm, come from SAPO-11's and alumina's nature, respectively. The introduction of active components has no significant effect on textural properties. The well-dispersed nanosized active phase of the prepared bifunctional catalysts was characterized by the HR TEM method to calculate morphological properties. The CoMoS active phase has slightly higher dispersion in comparison with the NiWS one.

CoMoS- and NiWS-catalysts demonstrate perfect stability during the experiments with sulfur-containing feedstock. The reaction rate constants are comparable for both sulfide catalysts during these experiments, and are equal to the initial values for pure n-C$_{16}$ processing; 0.8–0.9 × 10$^4$ mol g$^{-1}$ h$^{-1}$. Low sulfur concentration of 10 ppm in feedstock inhibits Pt-catalysts by about 50%. Regardless, Pt-catalyst provides more intense n-C$_{16}$ isomerization, and the reaction rate constant is two times higher in comparison with the NiWS-catalyst. The situation reverses when 100 ppm of sulfur is introduced into the feedstock. The inhibiting effect of sulfur for Pt-catalyst is 77% in this case, which is equal to an isomerization reaction rate constant 3.0 times lower than the NiWS-catalyst.

The catalytic behaviors of the prepared TMS- and Pt-based bifunctional catalysts are quite similar in nitrogen-containing feedstock processing. Isomerization activity is inhibited by nitrogen through a similar mechanism, which differs from the one for Pt-catalysts poisoned by sulfur. The residual catalytic activity is the result of the strong limitation of the isomerization reaction, caused by nitrogen in the feedstock. The isomerization reaction rate constants evaluated during the processing of the feedstock with 50 ppm of nitrogen increase from 0.09 to 0.21 × 10$^4$ mol g$^{-1}$ h$^{-1}$ μmol/g in the following order: Pt-, CoMoS-, NiWS-catalysts. In addition, nitrogen in the feedstock has a more intense inhibiting effect on isomerization catalytic activity. Reaching the steady state of n-hexadecane conversion takes about 20 h in the case of nitrogen poisoning, and about 38 h for sulfur-containing feedstock.

Therefore, it can be concluded that transition metal-based bifunctional catalysts have a significant advantage over Pt-catalyst in n-alkanes isomerization, when the amount of sulfur in the feedstock is equal to, or higher than, 100 ppm. Nitrogen has an intensive inhibiting effect on both TMS- and Pt-containing bifunctional catalysts, which is most probably due to support acid sites poisoning, in which case, additional improvements of the support are required.

**Supplementary Materials:** The following are available online at http://www.mdpi.com/2073-4344/10/6/594/s1. Figure S1. XRD spectra of synthesized $Al_2O_3$-SAPO-11 composite support; Figure S2. SEM image of $Al_2O_3$-SAPO-11 composite support; Figure S3. Chromatogram of the feedstock n-$C_{16}$ in hexane.

**Author Contributions:** Conceptualization, A.P. (Aleksey Pimerzin), A.P. (Andrey Pimerzin); methodology, A.M., A.S., A.V.; software, A.V., A.P. (Andrey Pimerzin); validation, A.P. (Aleksey Pimerzin), A.G.; formal analysis, A.S., A.M.; investigation, A.S., A.P. (Aleksey Pimerzin), A.P. (Andrey Pimerzin); resources, V.V., A.G.; data curation, A.P. (Aleksey Pimerzin), A.S.; writing—original draft preparation, A.S., A.M.; writing—review and editing, A.V., A.P. (Aleksey Pimerzin); visualization, A.V., A.S.; supervision, A.P. (Aleksey Pimerzin), A.G.; project administration, A.P. (Aleksey Pimerzin); funding acquisition A.P. (Aleksey Pimerzin), A.G. All authors have read and agreed to the published version of the manuscript.

**Funding:** This research was financially supported by the Russian Science Foundation (Project No. 19-79-00293).

**Conflicts of Interest:** The authors declare no conflict of interest.

## References

1. Mäki-Arvela, P.; Kaka khel, T.; Azkaar, M.; Engblom, S.; Murzin, D. Catalytic Hydroisomerization of Long-Chain Hydrocarbons for the Production of Fuels. *Catalysts* **2018**, *8*, 534. [CrossRef]
2. Len, C.; Luisi, R. Catalytic Methods in Flow Chemistry. *Catalysts* **2019**, *9*, 663. [CrossRef]
3. Lázaro, N.; Franco, A.; Ouyang, W.; Balu, A.; Romero, A.; Luque, R.; Pineda, A. Continuous-Flow Hydrogenation of Methyl Levulinate Promoted by Zr-Based Mesoporous Materials. *Catalysts* **2019**, *9*, 142. [CrossRef]
4. An, K.; Zhang, Q.; Alayoglu, S.; Musselwhite, N.; Shin, J.-Y.; Somorjai, G.A. High-Temperature Catalytic Reforming of n-Hexane over Supported and Core–Shell Pt Nanoparticle Catalysts: Role of Oxide–Metal Interface and Thermal Stability. *Nano Lett.* **2014**, *14*, 4907–4912. [CrossRef] [PubMed]
5. Beltramini, J.; Trimm, D.L. Catalytic reforming of n-heptane on platinum, tin and platinum-tin supported on alumina. *Appl. Catal.* **1987**, *31*, 113–118. [CrossRef]
6. Weyda, H.; Köhler, E. Modern refining concepts—An update on naphtha-isomerization to modern gasoline manufacture. *Catal. Today* **2003**, *81*, 51–55. [CrossRef]
7. Akhmedov, V.M.; Al-Khowaiter, S.H. Recent Advances and Future Aspects in the Selective Isomerization of High n-Alkanes. *Catal. Rev.* **2007**, *49*, 33–139. [CrossRef]
8. Guisnet, M.; Gilson, J.-P. *Zeolites for Cleaner Technologies Catalytic Science Series*; Hutchings, G.J., Ed.; Imperial College Press: London, UK, 2002; Volume 3, ISBN 1-86094-329-2.
9. Martens, J.A.; Verboekend, D.; Thomas, K.; Vanbutsele, G.; Gilson, J.-P.; Pérez-Ramírez, J. Hydroisomerization of Emerging Renewable Hydrocarbons using Hierarchical Pt/H-ZSM-22 Catalyst. *ChemSusChem* **2013**, *6*, 421–425. [CrossRef]
10. Mortier, R.M.; Fox, M.F.; Malcolm, F.; Orszulik, S.T. *Chemistry and Technology of Lubricants*; Springer: Berlin/Heidelberg, Germany, 2010; ISBN 9781402086625.
11. Klein, A.; Keisers, K.; Palkovits, R. Formation of 1,3-butadiene from ethanol in a two-step process using modified zeolite-β catalysts. *Appl. Catal. A Gen.* **2016**, *514*, 192–202. [CrossRef]
12. Mendes, P.S.F.; Mota, F.M.; Silva, J.M.; Ribeiro, M.F.; Daudin, A.; Bouchy, C. A systematic study on mixtures of Pt/zeolite as hydroisomerization catalysts. *Catal. Sci. Technol.* **2017**, *7*, 1095–1107. [CrossRef]
13. Guisnet, M. "Ideal" bifunctional catalysis over Pt-acid zeolites. *Catal. Today* **2013**, *218–219*, 123–134. [CrossRef]
14. Alvarez, F.; Ribeiro, F.R.; Perot, G.; Thomazeau, C.; Guisnet, M. Hydroisomerization and Hydrocracking of Alkanes: 7. Influence of the Balance between Acid and Hydrogenating Functions on the Transformation of n-Decane on PtHY Catalysts. *J. Catal.* **1996**, *162*, 179–189. [CrossRef]

15. Glotov, A.P.; Artemova, M.I.; Demikhova, N.R.; Smirnova, E.M.; Ivanov, E.V.; Gushchin, P.A.; Egazar'yants, S.V.; Vinokurov, V.A. A Study of Platinum Catalysts Based on Ordered Al–MCM-41 Aluminosilicate and Natural Halloysite Nanotubes in Xylene Isomerization. *Pet. Chem.* **2019**, *59*, 1226–1234. [CrossRef]
16. Radlik, M.; Śrębowata, A.; Juszczyk, W.; Matus, K.; Małolepszy, A.; Karpiński, Z. n-Hexane conversion on γ-alumina supported palladium–platinum catalysts. *Adsorption* **2019**, *25*, 843–853. [CrossRef]
17. van de Runstraat, A.; Kamp, J.A.; Stobbelaar, P.J.; van Grondelle, J.; Krijnen, S.; van Santen, R.A. Kinetics of Hydro-isomerization ofn-Hexane over Platinum Containing Zeolites. *J. Catal.* **1997**, *171*, 77–84. [CrossRef]
18. Suárez París, R.; L'Abbate, M.E.; Liotta, L.F.; Montes, V.; Barrientos, J.; Regali, F.; Aho, A.; Boutonnet, M.; Järås, S. Hydroconversion of paraffinic wax over platinum and palladium catalysts supported on silica–alumina. *Catal. Today* **2016**, *275*, 141–148. [CrossRef]
19. De Lucas, A.; Sánchez, P.; Fúnez, A.; Ramos, M.J.; Valverde, J.L. Influence of clay binder on the liquid phase hydroisomerization of n-octane over palladium-containing zeolite catalysts. *J. Mol. Catal. A Chem.* **2006**, *259*, 259–266. [CrossRef]
20. Wang, Y.; Tao, Z.; Wu, B.; Xu, J.; Huo, C.; Li, K.; Chen, H.; Yang, Y.; Li, Y. Effect of metal precursors on the performance of Pt/ZSM-22 catalysts for n-hexadecane hydroisomerization. *J. Catal.* **2015**, *322*, 1–13. [CrossRef]
21. Zhu, S.; Liu, S.; Zhang, H.; Lü, E.; Ren, J. Investigation of synthesis and hydroisomerization performance of SAPO-11/Beta composite molecular sieve. *Chin. J. Catal.* **2014**, *35*, 1676–1686. [CrossRef]
22. Karakhanov, E.A.; Glotov, A.P.; Nikiforova, A.G.; Vutolkina, A.V.; Ivanov, A.O.; Kardashev, S.V.; Maksimov, A.L.; Lysenko, S.V. Catalytic cracking additives based on mesoporous MCM-41 for sulfur removal. *Fuel Process. Technol.* **2016**, *153*, 50–57. [CrossRef]
23. Karakhanov, E.; Maximov, A.; Zolotukhina, A.; Vinokurov, V.; Ivanov, E.; Glotov, A. Manganese and Cobalt Doped Hierarchical Mesoporous Halloysite-Based Catalysts for Selective Oxidation of p-Xylene to Terephthalic Acid. *Catalysts* **2019**, *10*, 7. [CrossRef]
24. Glotov, A.; Stytsenko, V.; Artemova, M.; Kotelev, M.; Ivanov, E.; Gushchin, P.; Vinokurov, V. Hydroconversion of Aromatic Hydrocarbons over Bimetallic Catalysts. *Catalysts* **2019**, *9*, 384. [CrossRef]
25. Kramer, G.M.; Schriesheim, A. Heptane Isomerization Mechanism. *J. Phys. Chem.* **1961**, *65*, 1283–1286. [CrossRef]
26. Weitkamp, J. Hydrocracking, Cracking and Isomerization of Hydrocarbons.|Hydrocracken, cracken und isomerisieren von kohlenwasserstoffen. *Erdoel Kohle-Erdgas-Petrochem* **1978**, *31*, 13–22.
27. Krummenacher, J. Catalytic partial oxidation of higher hydrocarbons at millisecond contact times: Decane, hexadecane, and diesel fuel. *J. Catal.* **2003**, *215*, 332–343. [CrossRef]
28. Zhang, F.; Liu, Y.; Sun, Q.; Dai, Z.; Gies, H.; Wu, Q.; Pan, S.; Bian, C.; Tian, Z.; Meng, X.; et al. Design and preparation of efficient hydroisomerization catalysts by the formation of stable SAPO-11 molecular sieve nanosheets with 10–20 nm thickness and partially blocked acidic sites. *Chem. Commun.* **2017**, *53*, 4942–4945. [CrossRef]
29. Guo, L.; Fan, Y.; Bao, X.; Shi, G.; Liu, H. Two-stage surfactant-assisted crystallization for enhancing SAPO-11 acidity to improve n-octane di-branched isomerization. *J. Catal.* **2013**, *301*, 162–173. [CrossRef]
30. Deldari, H. Suitable catalysts for hydroisomerization of long-chain normal paraffins. *Appl. Catal. A Gen.* **2005**, *293*, 1–10. [CrossRef]
31. Bouchy, C.; Hastoy, G.; Guillon, E.; Martens, J.A. Fischer-Tropsch Waxes Upgrading via Hydrocracking and Selective Hydroisomerization. *Oil Gas Sci. Technol. Rev. l'IFP* **2009**, *64*, 91–112. [CrossRef]
32. Lakhapatri, S.L.; Abraham, M.A. Deactivation due to sulfur poisoning and carbon deposition on Rh-Ni/Al2O3 catalyst during steam reforming of sulfur-doped n-hexadecane. *Appl. Catal. A Gen.* **2009**, *364*, 113–121. [CrossRef]
33. Galperin, L.B. Hydroisomerization of N-decane in the presence of sulfur and nitrogen compounds. *Appl. Catal. A Gen.* **2001**, *209*, 257–268. [CrossRef]
34. Parmar, S.; Pant, K.K.; John, M.; Kumar, K.; Pai, S.M.; Newalkar, B.L. Hydroisomerization of n-hexadecane over Pt/ZSM-22 framework: Effect of divalent cation exchange. *J. Mol. Catal. A Chem.* **2015**, *404–405*, 47–56. [CrossRef]

35. Höchtl, M.; Jentys, A.; Vinek, H. Hydroisomerization of Heptane Isomers over Pd/SAPO Molecular Sieves: Influence of the Acid and Metal Site Concentration and the Transport Properties on the Activity and Selectivity. *J. Catal.* **2000**, *190*, 419–432. [CrossRef]
36. Geng, C.-H.; Zhang, F.; Gao, Z.-X.; Zhao, L.-F.; Zhou, J.-L. Hydroisomerization of n-tetradecane over Pt/SAPO-11 catalyst. *Catal. Today* **2004**, *93–95*, 485–491. [CrossRef]
37. Martens, J.A.; Verboekend, D.; Thomas, K.; Vanbutsele, G.; Pérez-Ramírez, J.; Gilson, J.-P. Hydroisomerization and hydrocracking of linear and multibranched long model alkanes on hierarchical Pt/ZSM-22 zeolite. *Catal. Today* **2013**, *218–219*, 135–142. [CrossRef]
38. Corma, A.; Martínez, A.; Martínez-Soria, V. Hydrogenation of Aromatics in Diesel Fuels on Pt/MCM-41 Catalysts. *J. Catal.* **1997**, *169*, 480–489. [CrossRef]
39. Yasuda, H.; Yoshimura, Y. Hydrogenation of tetralin over zeolite-supported Pd-Pt catalysts in the presence of dibenzothiophene. *Catal. Lett.* **1997**, *46*, 43–48. [CrossRef]
40. Escobar, J.; Núñez, S.; Montesinos-Castellanos, A.; de los Reyes, J.A.; Rodríguez, Y.; González, O.A. Dibenzothiophene hydrodesulfurization over PdPt/Al$_2$O$_3$–TiO$_2$. Influence of Ti-addition on hydrogenating properties. *Mater. Chem. Phys.* **2016**, *171*, 185–194. [CrossRef]
41. Xiong, J.; Ma, Y. Catalytic Hydrodechlorination of Chlorophenols in a Continuous Flow Pd/CNT-Ni Foam Micro Reactor Using Formic Acid as a Hydrogen Source. *Catalysts* **2019**, *9*, 77. [CrossRef]
42. Flego, C.; Galasso, L.; Vidotto, S.; Faraci, G. *Effects of H$_2$S on Bifunctional Catalysts*; Elsevier: Amsterdam, The Netherlands, 1997; pp. 479–486.
43. Marafi, M.; Stanislaus, A.; Furimsky, E. Catalyst Deactivation. In *Handbook of Spent Hydroprocessing Catalysts*; Elsevier: Amsterdam, The Netherlands, 2017; pp. 67–140.
44. Topsøe, H.; Clausen, B.S.; Massoth, F.E. *Hydrotreating Catalysis*; Springer: Berlin/Heidelberg, Germany, 1996.
45. Xing, G.; Liu, S.; Guan, Q.; Li, W. Investigation on hydroisomerization and hydrocracking of C 15 –C 18 n -alkanes utilizing a hollow tubular Ni-Mo/SAPO-11 catalyst with high selectivity of jet fuel. *Catal. Today* **2019**, *330*, 109–116.
46. Liu, P.; Wu, M.-Y.; Wang, J.; Zhang, W.-H.; Li, Y.-X. Hydroisomerization of n-heptane over MoP/Hβ catalyst doped with metal additive. *Fuel Process. Technol.* **2015**, *131*, 311–316. [CrossRef]
47. Karakoulia, S.A.; Heracleous, E.; Lappas, A.A. Mild hydroisomerization of heavy naphtha on mono- and bi-metallic Pt and Ni catalysts supported on Beta zeolite. *Catal. Today* **2019**. [CrossRef]
48. Stanislaus, A.; Marafi, A.; Rana, M.S. Recent advances in the science and technology of ultra low sulfur diesel (ULSD) production. *Catal. Today* **2010**, *153*, 1–68. [CrossRef]
49. van Veen, J.A.R. What's new? On the development of sulphidic HT catalysts before the molecular aspects. *Catal. Today* **2017**, *292*, 2–25. [CrossRef]
50. Vutolkina, A.V.; Makhmutov, D.F.; Zanina, A.V.; Maximov, A.L.; Glotov, A.P.; Sinikova, N.A.; Karakhanov, E.A. Hydrogenation of Aromatic Substrates over Dispersed Ni–Mo Sulfide Catalysts in System H$_2$O/CO. *Pet. Chem.* **2018**, *58*, 528–534. [CrossRef]
51. Vutolkina, A.V.; Makhmutov, D.F.; Zanina, A.V.; Maximov, A.L.; Kopitsin, D.S.; Glotov, A.P.; Egazar'yants, S.V.; Karakhanov, E.A. Hydroconversion of Thiophene Derivatives over Dispersed Ni–Mo Sulfide Catalysts. *Pet. Chem.* **2018**, *58*, 1227–1232. [CrossRef]
52. Han, W.; Nie, H.; Long, X.; Li, M.; Yang, Q.; Li, D. Effects of the support BrØnsted acidity on the hydrodesulfurization and hydrodenitrogention activity of sulfided NiMo/Al2O3catalysts. *Catal. Today* **2017**, *292*, 58–66.
53. Yu, Q.; Zhang, L.; Guo, R.; Sun, J.; Fu, W.; Tang, T.; Tang, T. Catalytic performance of CoMo catalysts supported on mesoporous ZSM-5 zeolite-alumina composites in the hydrodesulfurization of 4,6-dimethyldibenzothiophene. *Fuel Process. Technol.* **2017**, *159*, 76–87. [CrossRef]
54. Wang, X.; Mei, J.; Zhao, Z.; Zheng, P.; Chen, Z.; Gao, D.; Fu, J.; Fan, J.; Duan, A.; Xu, C. Self-Assembly of Hierarchically Porous ZSM-5/SBA-16 with Different Morphologies and Its High Isomerization Performance for Hydrodesulfurization of Dibenzothiophene and 4,6-Dimethyldibenzothiophene. *ACS Catal.* **2018**, *8*, 1891–1902. [CrossRef]
55. Mériaudeau, P.; Tuan, V.A.; Sapaly, G.; Nghiem, V.T.; Naccache, C. Pore size and crystal size effects on the selective hydroisomerisation of C8 paraffins over Pt–Pd/SAPO-11, Pt–Pd/SAPO-41 bifunctional catalysts. *Catal. Today* **1999**, *49*, 285–292. [CrossRef]

56. Yadav, R.; Sakthivel, A. Silicoaluminophosphate molecular sieves as potential catalysts for hydroisomerization of alkanes and alkenes. *Appl. Catal. A Gen.* **2014**, *481*, 143–160. [CrossRef]
57. Lee, E.; Yun, S.; Park, Y.-K.; Jeong, S.-Y.; Han, J.; Jeon, J.-K. Selective hydroisomerization of n-dodecane over platinum supported on SAPO-11. *J. Ind. Eng. Chem.* **2014**, *20*, 775–780. [CrossRef]
58. Kim, M.Y.; Lee, K.; Choi, M. Cooperative effects of secondary mesoporosity and acid site location in Pt/SAPO-11 on n-dodecane hydroisomerization selectivity. *J. Catal.* **2014**, *319*, 232–238. [CrossRef]
59. Pimerzin, A.A.; Roganov, A.A.; Verevkin, S.P.; Konnova, M.E.; Pilshchikov, V.A.; Pimerzin, A.A. Bifunctional catalysts with noble metals on composite Al2O3-SAPO-11 carrier and their comparison with CoMoS one in n-hexadecane hydroisomerization. *Catal. Today* **2019**, *329*, 71–81. [CrossRef]
60. Topsøe, H.; Clausen, B.S.; Topsøe, N.-Y.; Zeuthen, P. Progress in the Design of Hydrotreating Catalysts Based on Fundamental Molecular Insight. In *Studies in Surface Science and Catalysis*; Elsevier: Amsterdam, The Netherlands, 1989; Volume 53, pp. 77–102. ISBN 9780444882110.
61. Topsøe, H. The role of Co–Mo–S type structures in hydrotreating catalysts. *Appl. Catal. A Gen.* **2007**, *322*, 3–8. [CrossRef]
62. Pimerzin, A.A.; Savinov, A.A.; Ishutenko, D.I.; Verevkin, S.P.; Pimerzin, A.A. Isomerization of Linear Paraffin Hydrocarbons in the Presence of Sulfide CoMo and NiW Catalysts on $Al_2O_3$—SAPO-11 Support. *Russ. J. Appl. Chem.* **2019**, *92*, 1772–1779. [CrossRef]
63. Mériaudeau, P.; Tuan, V.A.; Nghiem, V.T.; Lai, S.Y.; Hung, L.N.; Naccache, C. SAPO-11, SAPO-31, and SAPO-41 Molecular Sieves: Synthesis, Characterization, and Catalytic Properties in n-Octane Hydroisomerization. *J. Catal.* **1997**, *169*, 55–66. [CrossRef]
64. Kasztelan, S.; Toulhoat, H.; Grimblot, J.; Bonnelle, J.P. A geometrical model of the active phase of hydrotreating catalysts. *Appl. Catal.* **1984**, *13*, 127–159. [CrossRef]
65. Hensen, E.J.; Kooyman, P.; van der Meer, Y.; van der Kraan, A.; de Beer, V.H.; van Veen, J.A.; van Santen, R. The Relation between Morphology and Hydrotreating Activity for Supported MoS2 Particles. *J. Catal.* **2001**, *199*, 224–235. [CrossRef]
66. Pimerzin, A.A.; Nikulshin, P.A.; Mozhaev, A.V.; Pimerzin, A.A.; Lyashenko, A.I. Investigation of spillover effect in hydrotreating catalysts based on $Co_2Mo_{10}$-heteropolyanion and cobalt sulphide species. *Appl. Catal. B Environ.* **2015**, *168–169*, 396–407. [CrossRef]

© 2020 by the authors. Licensee MDPI, Basel, Switzerland. This article is an open access article distributed under the terms and conditions of the Creative Commons Attribution (CC BY) license (http://creativecommons.org/licenses/by/4.0/).

*Communication*

# Coupling Pre-Reforming and Partial Oxidation for LPG Conversion to Syngas

**Dmitriy I. Potemkin** [1,2,3,*], **Vladimir N. Rogozhnikov** [3,4], **Sergey I. Uskov** [2,3], **Vladislav A. Shilov** [2,3], **Pavel V. Snytnikov** [2,3] and **Vladimir A. Sobyanin** [2,3]

1. Department of Environmental Engineering, Novosibirsk State Technical University, Karl Marx Pr., 20, 630073 Novosibirsk, Russia
2. Department of Natural Sciences, Novosibirsk State University, Pirogova St., 2, 630090 Novosibirsk, Russia; serg5810@gmail.com (S.I.U.); dubbmx97@gmail.com (V.A.S.); pvsnyt@catalysis.ru (P.V.S.); sobyanin@catalysis.ru (V.A.S.)
3. Boreskov Institute of Catalysis, Pr. Lavrentieva, 5, 630090 Novosibirsk, Russia; rvn@catalysis.ru
4. Faculty of Oil and Gas Field Development, Gubkin Russian State University of Oil and Gas, Leninsky Pr., 65, 119991 Moscow, Russia
* Correspondence: potema@catalysis.ru; Tel.: +7-913-932-46-20

Received: 31 August 2020; Accepted: 18 September 2020; Published: 21 September 2020

**Abstract:** Coupling of the pre-reforming and partial oxidation was considered for the conversion of liquefied petroleum gas to syngas for the feeding applications of solid oxide fuel cells. Compared with conventional two step steam reforming, it allows the amount of water required for the process, and therefore the energy needed for water evaporation, to be lowered; substitution of high-potential heat by lower ones; and substitution of expensive tubular steam reforming reactors by adiabatic ones. The supposed process is more productive due to the high reaction rate of partial oxidation. The obtained syngas contains only ca. 10 vol.% $H_2O$ and ca. 50 vol.% of $H_2 + CO$, which is attractive for the feeding application of solid oxide fuel cells. Compared with direct partial oxidation of liquefied petroleum gas, the suggested scheme is more energy efficient and overcomes problems with coke formation and catalyst overheating. The proof-of-concept experiments were carried out. The granular Ni-$Cr_2O_3$-$Al_2O_3$ catalyst was shown to be effective for propane pre-reforming at 350–400 °C, $H_2O$:C molar ratio of 1.0, and flow rate of 12,000 $h^{-1}$. The composite Rh/$Ce_{0.75}Zr_{0.25}O_{2-\delta}$-$\eta$-$Al_2O_3$/FeCrAl catalyst was shown to be active and stable under conditions of partial oxidation of methane-rich syngas after pre-reforming and provided a syngas ($H_2 + CO$) productivity of 28 $m^3 \cdot L_{cat}^{-1} \cdot h^{-1}$ (standard temperature and pressure).

**Keywords:** syngas; pre-reforming; partial oxidation; tri-reforming; LPG; propane; SOFC

## 1. Introduction

The solid oxide fuel cell (SOFC) is an attractive device for direct conversion of chemical energy of fuel to electricity with additional heat production [1]. Due to its high working temperature, a SOFC can operate on both pure hydrogen and synthesis gas (syngas) [2] obtained from hydrocarbon fuels (methane, liquified petroleum gas (LPG), liquid hydrocarbons) by steam conversion [3], partial oxidation (PO) [4], dry reforming [5,6], or autothermal reforming [7]. In the case of a SOFC stack with power of several kilowatts or more, the heat released during the reaction can be efficiently utilized to conduct endothermal steam reforming of the initial fuel. This increases the efficiency of the entire system (reformer + SOFC system). For small power SOFCs, usually less than 1–2 kW, the heat generated is not always enough to simultaneously maintain the required stack temperature and for endothermic steam reforming. In this case, it is preferable to use an exothermic partial oxidation process:

$$2CH_4 + O_2 \rightarrow 2CO + 4H_2 \ (\Delta_r H^0_{973K} = -46.4 \text{ kJ/mol})$$

However, the overall efficiency of the power plant is significantly reduced. In addition, when using LPG (which can be compactly stored in a liquid state in cylinders at low pressure) during partial oxidation, the conversion temperature usually exceeds 900 °C, which leads to accelerated degradation of the catalyst [8].

One of the options for increasing the efficiency of fuel conversion is a process that combines the weakly exothermic process of LPG pre-reforming at low temperatures and partial oxidation of the resulting methane–hydrogen mixture, the temperature of which is significantly lower than the temperature of LPG partial oxidation:

$$4C_nH_{2n+2} + 2(n-1)H_2O \rightarrow (3n+1)CH_4 + (n-1)CO_2 \ (\Delta_r H^0_{673K} = -80.8 \text{ kJ/mol for } n = 3)$$

In addition, the pre-reforming process requires much smaller amounts of water to be supplied than for the classic steam reforming process [9].

In this work we considered the possibility of combining pre-reforming at a low steam to carbon ratio and subsequent partial oxidation to maximize the energy efficiency of syngas production from LPG and carried out proof-of-concept experiments.

## 2. Results and Discussion

Propane, the main component of LPG, was chosen as a model compound. In our previous work the Ni-Cr$_2$O$_3$-Al$_2$O$_3$ catalyst was shown to be effective in propane pre-reforming [10]. As an example, Figure 1 shows the temperature dependences of the outlet concentrations of propane, methane, carbon dioxide, and hydrogen during propane pre-reforming under 1 bar pressure, with a H$_2$O:C molar ratio of 1.0 and gas hourly space velocity (GHSV) of 12,000 h$^{-1}$. It is seen that the propane conversion and CH$_4$, H$_2$ and CO$_2$ concentrations increased with the increase of temperature, reaching equilibrium values at 380 °C.

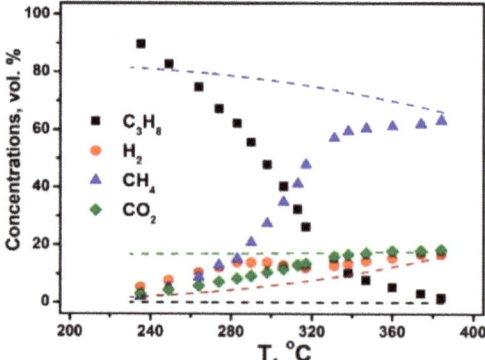

**Figure 1.** The outlet concentrations of propane, methane, carbon dioxide, and hydrogen (on the dry basis) as a function of temperature during propane pre-reforming. Experimental conditions: 240–400 °C, GHSV 12,000 h$^{-1}$, 1 bar pressure, reaction mixture: 25 vol.% C$_3$H$_8$, 75 vol.% H$_2$O. Points are experimental. Dashed lines are equilibrium concentrations.

Methane-rich gas (MRG) after pre-reforming typically contains 40–70 vol.% CH$_4$, 10–30 vol.% H$_2$O, 5–15 vol.% CO$_2$, 5–20 vol.% H$_2$, and 0.1–2 vol.% CO. The obtained methane-rich gas can be converted to syngas via the addition of the required quantity of oxygen or air. As the resulting mixture will contain water steam together with oxygen, the process will be near autothermal, which is especially convenient for compact systems.

It was previously demonstrated [8,11] that Rh/Ce$_{0.75}$Zr$_{0.25}$O$_{2-\delta}$–η-Al$_2$O$_3$/FeCrAl catalyst (further Rh-block) was highly efficient in the autothermal reforming of diesel fuel, exhibiting high catalytic activity, stability, and a low tendency to carbonization. In this work, its properties in the partial oxidation of methane-rich gas are studied.

Figure 2 shows the dependences of the outlet concentrations of $H_2$, $N_2$, $CO$, $CO_2$, and $CH_4$ and the temperatures at the inlet ($T_{in}$) and outlet ($T_{out}$) centerline points of the catalytic block as a function of the flow rate of the reaction mixture (GHSV) for the PO of MRG over the Rh-block at 730 °C. It is seen that, for all flows, the distribution of products was close to equilibrium at 700 °C. The temperature at the outlet of the block varied from 693 to 730 °C, i.e., was close to 700 °C, which explains the closeness of the product distribution to equilibrium. However slight increase of $CH_4$ outlet concentration, accompanied with the decrease of $H_2$ and $CO$ level, was observed at GHSVs 40,000–60,000 h$^{-1}$, indicating the limitation of the overall reaction rate.

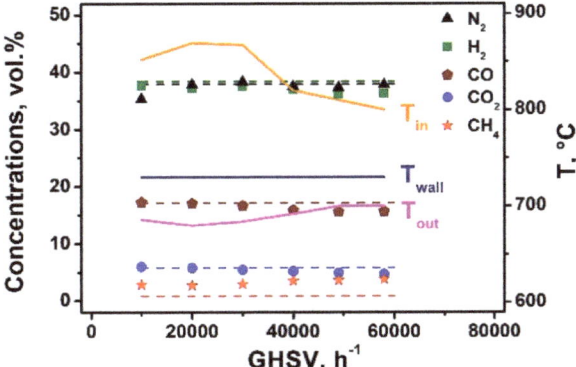

**Figure 2.** The inlet ($T_{in}$) and outlet ($T_{out}$) temperatures of the Rh-block, $H_2$, $N_2$, $CO$, $CO_2$, and $CH_4$ concentrations (on a dry basis) as a function of space velocity (GHSV) for partial oxidation (PO) of methane-rich gas (MRG). Experimental conditions: GHSV 18,000–55,000 h$^{-1}$, 1 bar pressure, reactor wall temperature 730 °C, reaction mixture (vol.%): 25.6 $CH_4$, 13.4 $O_2$, 10.5 $H_2O$, and 50.5 $N_2$. Points are experimental. Dashed lines are equilibrium concentrations.

The temperature at the catalyst inlet exceeded the reactor wall temperature (730 °C) and the outlet temperature, which was associated with the exothermic reaction of complete oxidation of methane and hydrogen. The outlet temperature, on the contrary, did not exceed the temperature of the walls of the reactor, which was due to the endothermic reactions of steam and dry reforming of methane. In this case, with an increase in the flow rate, $T_{in}$ decreased and $T_{out}$ increased. This is apparently associated with an increase in the heat release rate on the catalyst and with the stretching of the zone in which complete oxidation of methane and hydrogen occurs, with an increase of GHSV, which is caused by external diffusion limitation of the rate of total oxidation [12].

The stability of catalytic properties was studied during 25 h on stream. The test was performed at a reactor wall temperature of 730 °C and GHSV of 20,000 h$^{-1}$ and the following reaction mixture (vol.%): 25.6 $CH_4$, 13.4 $O_2$, 10.5 $H_2O$, and 50.5 $N_2$. The GHSV was periodically raised to 55,000 h$^{-1}$ (Figure 3a) to check the catalyst activity in the conditions at which the reaction was under kinetic control (product distribution differs from the equilibrium). The properties of the catalyst were constant during 25 h on stream (Figure 3b). The weight of the block did not change after the experiment; no damage (detachments) of the catalytic coating was detected. No carbon deposits were detected by temperature-programmed oxidation. Thus, the Rh-block showed stable operation under PO MRG conditions and provided a syngas ($H_2$ + CO) productivity of 28 $m^3 \cdot L_{cat}^{-1} \cdot h^{-1}$ (STP).

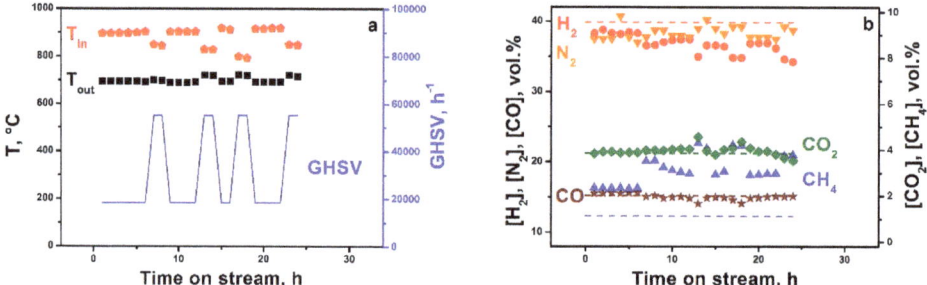

**Figure 3.** The inlet ($T_{in}$) and outlet ($T_{out}$) temperatures of the Rh-block and space velocity (**a**); $H_2$, $N_2$, CO, $CO_2$, and $CH_4$ concentrations (**b**) (on a dry basis) as a function of time on stream for PO of methane-rich syngas. Experimental conditions: GHSV 18,000–55,000 $h^{-1}$, 1 bar pressure, reactor wall temperature 730 °C, reaction mixture (vol.%): 25.6 $CH_4$, 13.4 $O_2$, 10.5 $H_2O$, and 50.5 $N_2$. Points are experimental. Dashed lines are equilibrium concentrations.

We also studied the light-on curve for the Rh-block in PO MRG, as it is important to determine the minimal outlet temperature after pre-reforming at which the PO could be launched. Figure 4 shows the dependence of the temperatures at the inlet ($T_{in}$) and outlet ($T_{out}$) centerline points of the catalytic block and the outlet concentrations of $H_2$, $N_2$, CO, $CO_2$, and $CH_4$ as a function of the reactor wall temperature. It is seen that at the initial temperature of 350 °C, $T_{in}$ and $T_{out}$ were similar to $T_{wall}$, indicating the absence of a PO reaction. A $T_{wall}$ increase to 400 °C induced a sharp rise of $T_{in}$ and $T_{out}$ to 598 and 465 °C, respectively. Thus, the temperature of 400 °C could be considered as a light-on one. Further increases of $T_{wall}$ provided a gradual increase of $T_{in}$, $T_{out}$, and progress of the reaction. Product distribution at 600–730 °C was close to equilibrium. Rapid product composition changes at 400–600 °C were difficult to register by gas chromatography. Transient switch on/off regimes will be the subject of further studies.

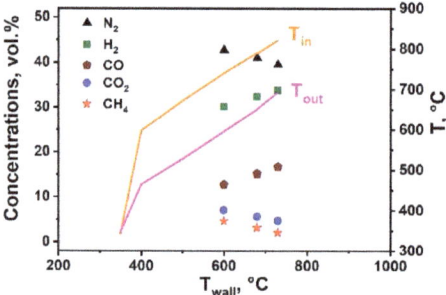

**Figure 4.** The inlet ($T_{in}$) and outlet ($T_{out}$) temperatures of the Rh-block and $H_2$, $N_2$, CO, $CO_2$, and $CH_4$ concentrations (on a dry basis) as a function of reactor wall temperature ($T_{wall}$) for PO of MRG. Experimental conditions: GHSV 11,000 $h^{-1}$, 1 bar pressure, reaction mixture (vol.%): 21.5 $CH_4$, 11.3 $O_2$, 8.8 $H_2O$, 5 $CO_2$, and 53.4 $N_2$.

Coupling pre-reforming and partial oxidation of the following syngas production scheme could be supposed (Figure 5): after mixing and heating the mixture of LPG and water steam ($H_2O$:C = 1, molar) was supplied to the adiabatic pre-reformer at 400 °C; the outlet temperature of methane-rich gas (MRG) was 410 °C; heated air was mixed with MRG and supplied to the adiabatic partial oxidation reactor at

410 °C; the resulted syngas with temperature of 670 °C was supplied directly to SOFC. The obtained syngas contained (vol.%) 1.1 $CH_4$, 36.4 $H_2$, 13.1 CO, 10.7 $H_2O$, 7.1 $CO_2$, 0.4 Ar, and 31.2 $N_2$ and could be successfully used for SOFC feeding. For 2 $kW_e$ SOFC only 0.2 $Nm^3$/h LPG (80 mol.% $C_3H_8$ and 20 mol.% $C_4H_{10}$) was required due to the high energy density of LPG. Furthermore, only 0.75 kW of heat power was required for water evaporation and gas superheating. Instead of the conventional steam reforming process, the heat in this scheme had low- and medium-potential: the maximal temperature had to be reached at 410 °C. Thus, the process needs could be satisfied by SOFC heat output (ca. 0.5 kW for 2 $kW_e$ SOFC) and anode gases afterburning (ca. 0.4 kW).

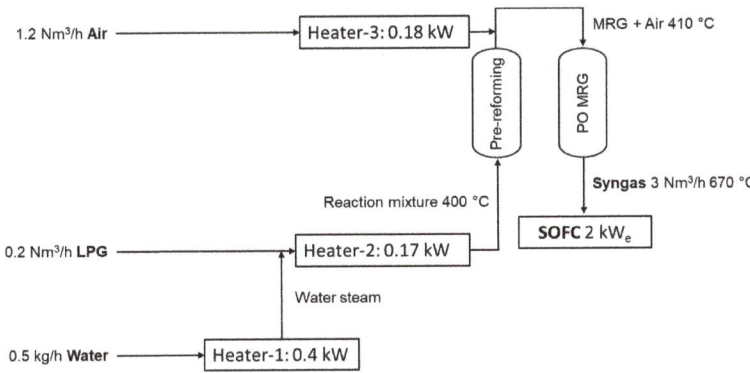

**Figure 5.** General scheme for the process of liquified petroleum gas (LPG) conversion to syngas for Figure 2. $kW_e$ solid oxide fuel cell (SOFC).

## 3. Conclusions

Compared with conventional two step steam reforming, coupling the pre-reforming and partial oxidation provides the reduction of amount of the water required for the process and therefore energy needed for its evaporation, substitution of high-potential heat by lower ones and expensive tubular steam reforming reactor by adiabatic one. In addition, the supposed process is more productive due to the high reaction rate of partial oxidation. The obtained syngas contains only ca. 10 vol.% $H_2O$ and ca. 50 vol.% of $H_2$ + CO which is attractive for SOFC feeding application. Compared with direct partial oxidation of LPG, the suggested scheme is more energy efficient and overcomes problems with coke formation and catalyst overheating.

## 4. Materials and Methods

Propane pre-reforming was studied using NIAP-07-05 industrial catalyst (wt.%): 42 NiO, 12 $Cr_2O_3$, 46 $Al_2O_3$, and 4 graphite (NIAP Ltd., Novomoskovsk, Russia). This catalyst demonstrated high activity in the low-temperature steam conversion of light hydrocarbons in methane excess [9,10].

The experiments on propane pre-reforming were carried out in a fixed-bed U-shaped quartz reactor (internal diameter 8 mm, catalyst bed volume 1.7 $cm^3$). Experimental conditions were as follows: 1 bar pressure, temperature of 240–400 °C, GHSV of 12,000 $h^{-1}$. Before the experiment, 2 g of the catalyst (fraction 0.5–1 mm) were reduced at 420 °C in a flow of 10 mL/min $H_2$ and 100 mL/min Ar. The composition of the initial reaction mixture was 25 vol.% $C_3H_8$ and 75 vol.% $H_2O$, which corresponded to an $H_2O$/C molar ratio of 1. The temperature of the catalyst was measured using a K-type thermocouple placed in the center of the catalyst bed. The composition of the gas phase components were determined using a Chromos GC-1000 chromatograph (CHROMOS Engineering company, Moscow, Russia). The concentration of carbon monoxide was negligible (below 0.1 vol.%), and therefore CO was not taken into account when processing the results. The carbon imbalance in all the experiments did not exceed ± 1 rel.%.

The composite Rh/Ce$_{0.75}$Zr$_{0.25}$O$_{2-\delta}$–η-Al$_2$O$_3$/FeCrAl (Rh-block) was prepared according to [8]. The cylindric Rh-block was 60 mm long and 18 mm in diameter. The weight of the Rh-block was 13.7 g, Total weight of the Rh/ Ce$_{0.75}$Zr$_{0.25}$O$_{2-\delta}$–η-Al$_2$O$_3$ coating was 1.58 g.

Partial oxidation of methane-rich gas over the Rh/Ce$_{0.75}$Zr$_{0.25}$O$_{2-\delta}$–η-Al$_2$O$_3$/FeCrAl catalyst was studied in a stainless steel flow reactor under atmospheric pressure. The catalyst was preliminarily reduced in a flow of 5 vol.% H$_2$/N$_2$ at 600 °C for 1 h. Afterwards, the reaction mixture was fed into the reactor. The volumetric flow rate of the mixture was varied in the range 10,000–55,000 h$^{-1}$. The temperature at the inlet and outlet of the block was measured with thermocouples along the central axis of the block. The composition of the gas phase components of the reaction mixture was determined using a Chromos GC-1000 (CHROMOS Engineering company, Moscow, Russia) gas chromatograph equipped with a thermal conductivity detector (TCD) and a flame ionization detector (FID) with a methanator. Before analysis, water was removed from the reformate using a moisture trap. Concentrations of H$_2$ and N$_2$ separated in a CaA column with an Ar carrier gas were determined with TCD. The CO, CO$_2$, and CH$_4$ were separated in a Chromosorb 106 column and quantified using FID. The "blank" experiment without a catalyst showed that at the used volumetric flow rates, the reaction on the walls of the reactor was not significant and could be ignored when analyzing the results.

Equilibrium concentrations were calculated using the HSC Chemistry 7.1 software package (HSC Chemistry, version 7.1, Outotec (company), 2011) under the assumption that the equilibrium mixture contains only gaseous substances.

**Author Contributions:** Conceptualization, D.I.P.; methodology, D.I.P.; software, validation, P.V.S.; formal analysis, investigation, D.I.P., S.I.U., V.A.S. (Vladislav A. Shilov), V.N.R.; resources, data curation, writing—original draft preparation, D.I.P.; supervision, V.A.S. (Vladimir A. Sobyanin). All authors have read and agreed to the published version of the manuscript.

**Funding:** The work is supported by the Russian Foundation for Basic Research under the project 19-33-60008 "Perspectiva".

**Conflicts of Interest:** The authors declare no conflict of interest.

## References

1. Ormerod, R.M. Solid oxide fuel cells. *Chem. Soc. Rev.* **2003**, *32*, 17–28. [CrossRef] [PubMed]
2. Amiri, A.; Tang, S.; Steinberger-Wilckens, R.; Tadé, M.O. Evaluation of fuel diversity in Solid Oxide Fuel Cell system. *Int. J. Hydrog. Energy* **2018**, *43*, 23475–23487. [CrossRef]
3. Abatzoglou, N.; Fauteux-Lefebvre, C. Review of catalytic syngas production through steam or dry reforming and partial oxidation of studied liquid compounds. *Wiley Interdiscip. Rev. Energy Environ.* **2016**, *5*, 169–187. [CrossRef]
4. Choudhary, T.V.; Choudhary, V.R. Energy-Efficient Syngas Production through Catalytic Oxy-Methane Reforming Reactions. *Angew. Chemie Int. Ed.* **2008**, *47*, 1828–1847. [CrossRef] [PubMed]
5. Sudhakaran, M.S.P.; Hossain, M.; Gnanasekaran, G.; Mok, Y. Dry Reforming of Propane over γ-Al2O3 and Nickel Foam Supported Novel SrNiO3 Perovskite Catalyst. *Catalysts* **2019**, *9*, 68. [CrossRef]
6. Råberg, L.B.; Jensen, M.B.; Olsbye, U.; Daniel, C.; Haag, S.; Mirodatos, C.; Sjåstad, A.O. Propane dry reforming to synthesis gas over Ni-based catalysts: Influence of support and operating parameters on catalyst activity and stability. *J. Catal.* **2007**, *249*, 250–260. [CrossRef]
7. Shoynkhorova, T.B.; Rogozhnikov, V.N.; Simonov, P.A.; Snytnikov, P.V.; Salanov, A.N.; Kulikov, A.V.; Gerasimov, E.Y.; Belyaev, V.D.; Potemkin, D.I.; Sobyanin, V.A. Highly dispersed Rh/Ce0.75Zr0.25O2-δ-η-Al2O3/FeCrAl wire mesh catalyst for autothermal n-hexadecane reforming. *Mater. Lett.* **2018**, *214*, 290–292. [CrossRef]
8. Shoynkhorova, T.B.; Rogozhnikov, V.N.; Ruban, N.V.; Shilov, V.A.; Potemkin, D.I.; Simonov, P.A.; Belyaev, V.D.; Snytnikov, P.V.; Sobyanin, V.A. Composite Rh/Zr0.25Ce0.75O2-Δ-η-Al2O3/Fecralloy wire mesh honeycomb module for natural gas, LPG and diesel catalytic conversion to syngas. *Int. J. Hydrog. Energy* **2019**, *44*, 9941–9948. [CrossRef]
9. Uskov, S.I.; Potemkin, D.I.; Shigarov, A.B.; Snytnikov, P.V.; Kirillov, V.A.; Sobyanin, V.A. Low-temperature steam conversion of flare gases for various applications. *Chem. Eng. J.* **2019**, *368*, 533–540. [CrossRef]

10. Uskov, S.I.; Potemkin, D.I.; Enikeeva, L.V.; Snytnikov, P.V.; Gubaydullin, I.M.; Sobyanin, V.A. Propane Pre-Reforming into Methane-Rich Gas over Ni Catalyst: Experiment and Kinetics Elucidation via Genetic Algorithm. *Energies* **2020**, *13*, 3393. [CrossRef]
11. Shoynkhorova, T.B.; Simonov, P.A.; Potemkin, D.I.; Snytnikov, P.V.; Belyaev, V.D. Applied Catalysis B: Environmental sorption-hydrolytic deposition for diesel fuel reforming to syngas. *Appl. Catal. B Environ.* **2018**, *237*, 237–244. [CrossRef]
12. Kirillov, V.A.; Shigarov, A.B.; Kuzin, N.A.; Kireenkov, V.V.; Brayko, A.S.; Burtsev, N.V. Ni/MgO Catalysts on Structured Metal Supports for the Air Conversion of Low Alkanes into Synthesis Gas. *Catal. Ind.* **2020**, *12*, 66–76. [CrossRef]

© 2020 by the authors. Licensee MDPI, Basel, Switzerland. This article is an open access article distributed under the terms and conditions of the Creative Commons Attribution (CC BY) license (http://creativecommons.org/licenses/by/4.0/).

Article

# Selective Hydrogenation of Acetylene over Pd-Mn/Al$_2$O$_3$ Catalysts

Dmitry Melnikov *, Valentine Stytsenko, Elena Saveleva, Mikhail Kotelev, Valentina Lyubimenko, Evgenii Ivanov, Aleksandr Glotov and Vladimir Vinokurov

Physical and Colloid Chemistry Department, Gubkin Russian State University of Oil and Gas, 65 Leninsky prosp, 119991 Moscow, Russia; vds41@mail.ru (V.S.); savel0394@gmail.com (E.S.); kain@inbox.ru (M.K.); ljubimenko@mail.ru (V.L.); ivanov166@list.ru (E.I.); glotov.a@gubkin.ru (A.G.); vinok_ac@mail.ru (V.V.)
* Correspondence: melnikov.dp@mail.ru

Received: 17 April 2020; Accepted: 2 June 2020; Published: 4 June 2020

**Abstract:** Novel bimetallic Pd-Mn/Al$_2$O$_3$ catalysts are designed by the decomposition of cyclopentadienylmanganese tricarbonyl (cymantrene) on reduced Pd/Al$_2$O$_3$ in an H$_2$ atmosphere. The peculiarities of cymantrene decomposition on palladium and, thus, the formation of bimetallic Pd-Mn catalysts are studied. The catalysts are characterized by N$_2$ adsorption, H$_2$ pulse chemisorption, temperature-programmed desorption of hydrogen (TPD-H$_2$), transmission electron microscopy (TEM), energy-dispersive X-ray spectroscopy (EDX), X-ray diffraction (XRD), and diffuse reflectance infrared Fourier transform spectroscopy (DRIFTS). The modified catalysts show the changed hydrogen chemisorption properties and the absence of weakly bonded hydrogen. Using an organomanganese precursor provides an uniform Mn distribution on the catalyst surface. Tested in hydrogenation of acetylene, the catalysts show both higher activity and selectivity to ethylene (20% higher) compared to the non-modified Pd/Al$_2$O$_3$ catalyst. The influence of the addition of Mn and temperature treatment on catalyst performance is studied. The optimal Mn content and treatment temperature are found. It is established that modification with Mn changes the route of acetylene hydrogenation from a consecutive scheme for Pd/Al$_2$O$_3$ to parallel one for the Pd-Mn samples. The reaction rate shows zero overall order by reagents for all tested catalysts.

**Keywords:** acetylene hydrogenation; ethylene production; bimetallic catalysts; palladium; manganese; cymantrene

## 1. Introduction

Ethylene is one of the commonly used monomers in the petrochemical industry worldwide and is produced by the steam cracking of hydrocarbons. Ethylene cuts typically comprise 0.5%–2% of acetylene, which is a poison for the polymerization catalysts and should be removed by selective hydrogenation to ethylene [1]. A number of active metals (Pd, Ni, Au) modified with a wide range of elements (Ag, Cu, Si, Ga, Sn, Pb, In, S, Fe) and supported on various carriers (Al$_2$O$_3$, SiO$_2$, TiO$_2$, ZnO) were investigated [2–25].

Monometallic Pd catalysts show a high activity but low selectivity to ethylene, so Pd is typically promoted with other metals. In industry, Pd-Ag/Al$_2$O$_3$ catalysts are widely used and much research is devoted to Pd-Ag compositions supported on alumina or silica. It is supposed that the promotion is based on an increased electronic density of the Pd *d*-band resulting in a decrease in ethylene [3] or hydrogen adsorption with further spill over [4]. In addition, it is suggested that the promotion is caused by not only an electronic but also a geometric effect [5], or just geometric [8]. Pd-Ag catalysts expose not only a higher selectivity to ethylene, but also a lower yield of C$_{6+}$ hydrocarbons (green oil) as compared to Pd/Al$_2$O$_3$, which is crucial for the cycle length of the catalysts [9]. However, problems with the deactivation of Pd-Ag catalysts during the selective hydrogenation of acetylene are as actual as before [8]. The main drawback of promotion with Ag is the significant reduction in catalyst activity—about 20 times as low as pure Pd [12]. Pure nickel shows a lower selectivity than even pure Pd, but the addition of Zn (such as Ni-Zn/MgAl$_2$O$_4$) increases the selectivity to the level of Pd-Ag catalysts [17] and decreases oligomerization [26].

Besides Ag, a number of other metals were investigated as promotors. The addition of Ga leads to an increased selectivity (71% at 99% conversion) compared to Pd catalysts, but also with Pd-Ag (49% at 83% conversion) [11–13]. The increased selectivity is explained by the isolation of the Pd sites [11,12] and the additional modification of the Fermi level of palladium [13]. The activity of Pd-Ga catalysts is similar to that of Pd-Ag.

Palladium modification with Cu shows a benefit in selectivity compared to Pd-Ag/Al$_2$O$_3$ [15,16] only when it is provided by the surface redox method, which is explained by blocking low coordinated Pd atoms (responsible for the low selectivity to ethylene) and by the hydrogenation properties of copper. Pure copper, however, requires significantly higher operating temperatures and shows an unacceptably high oligomer yield (up to 40%) [19].

A Pd-Zn catalyst supported on carbon or Al$_2$O$_3$ also shows a higher selectivity (+20%–50%) compared to pure palladium [20,21]. It was previously proved that Pd and Zn form a nanoalloy [27]. Moreover, Zn decreases the acidity of support and, hence, green oil formation.

As the carriers for hydrogenation catalysts, natural clay nanotubes such as halloysite are of particular interest [28–30]. Halloysite has the appropriate surface area (50–300 m$^2$.g$^{-1}$), a high ion-exchange capacity, and a micro-mesoporous structure that enables the synthesis of highly active catalysts and new materials applied for heterogeneous catalytic systems. Thus, a new approach was developed—a self-assembling synthesis of structured mesoporous silica on clay nanotubes (HNT), which was applied to create the highly porous material MCM-41-HNT with an enhanced thermal and mechanical stability [31].

Hard reducible oxides, such as Ce, Ti, and Nb, are also investigated as promotors [32,33]. The most efficient was TiO$_2$, however the catalyst selectivity did not exceed 50% at 90% conversion. The promotion effect is explained by the geometric and electronic modification of the Pd surface.

Supported on glass nanofibers, Pd also shows a high selectivity (up to ~56% at total conversion) [34,35]. The high selectivity is explained by: 1) the stabilized small Pd particles (~1 nm) in the subsurface of the glass fibers and 2) the much higher adsorption ability of acetylene compared to ethylene on Pd inside a glass matrix. As a result, the hydrogenation of ethylene from the gas phase is actually absent. Besides palladium, another interesting active metal in acetylene hydrogenation is gold. It is reported that Au/Al$_2$O$_3$ shows 100% selectivity at temperatures of 313–523 K, because ethylene hydrogenation only starts at temperatures above 573 K [23]. A lower selectivity was achieved on Au/TiO$_2$ (90% at 88% conversion). Au-Pd/TiO$_2$ catalysts show a higher activity compared to Au/TiO$_2$, but their selectivity is lower [24].

The addition of iron in the form of Fe$^0$ to Pd increases the selectivity to the olefin in the hydrogenation of both acetylene (88% at 87% conversion) [22] and phenylacetylene (90% at 99% conversion) [36,37].

The preparation of bimetallic catalysts (BMC) comprising VIII group metals by the decomposition of organometallic compounds has been patented [38–40] and reviewed in [41]. The decomposition of organometallic species under reduction conditions enables an easy formation of bimetallic catalysts with a zero valence state of the second metal. Some examples of BMC having unusual properties are as follows: Rh-Sn (butyl) [42], Pd-Pb (butyl) [43], Ni-Cr (arene) [44], and Pd-Fe (ferrocene) [22].

In this study, a number of Pd-Mn/Al$_2$O$_3$ catalysts were prepared by the decomposition of cymantrene on a reduced Pd/Al$_2$O$_3$ precursor. The use of cymantrene has some peculiarities as the molecule contains two types of ligands: CO and cyclopentadienyl. The catalysts were tested in a selective hydrogenation of acetylene. In all cases, Mn increases the catalyst's selectivity to ethylene as compared with the Pd/Al$_2$O$_3$ sample. Moreover, the Mn-modified samples have shown higher activity. It is found that the addition of Mn suppresses the hydrogen chemisorption on Pd catalysts.

## 2. Results and Discussion

### 2.1. Cymantrene Decomposition on Pd/Al$_2$O$_3$

To investigate the formation of Pd-Mn/Al$_2$O$_3$ catalysts, the decomposition of cymantrene on Pd/Al$_2$O$_3$ was performed in a temperature-programmed regime in an H$_2$ flow with a mass spectrometry analysis of effluent gas. Figure 1 shows the mass-spectra of cymantrene decomposition products in the range of 40–400 °C.

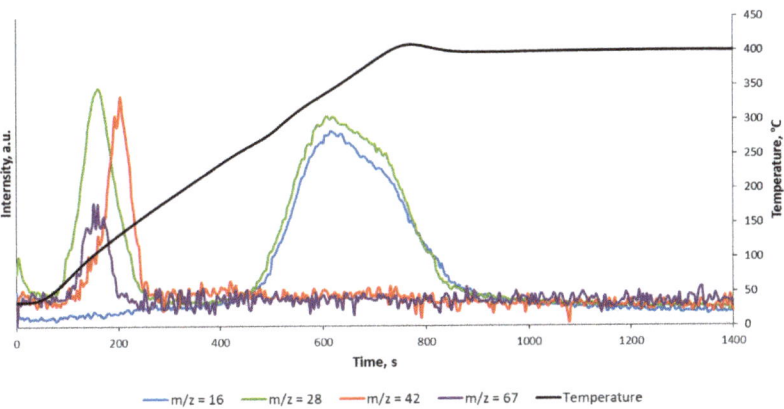

**Figure 1.** Decomposition of cymantrene on 0.068%Pd/Al$_2$O$_3$ in H$_2$ flow at temperatures of 40–400 °C.

As shown, at the initial step (temperatures of 80–150 °C) there are peaks with m/z 28 (carbon monoxide), 42 (cyclopentane), and 67 (cyclopentene) [45]. In the temperature range of 270–400 °C, one can observe two peaks with m/z 16 (methane) and 28 (carbon monoxide). The peaks corresponding to cyclopentadiene (m/z 65, 66) are absent. More details about the mass spectra interpretation are shown in the Supplementary Materials.

We may conclude, therefore, that the cyclopentadienyl ligand of cymantrene is removed after hydrogenation, mainly as cyclopentane at 80–150 °C. As for carbon monoxide, it is strongly bonded with metals, and may be removed as methane at temperatures above 270 °C [46]. However, as evidenced by the mass-spectrometric analysis of the effluent gas, to complete a CO removal a treatment in an H$_2$ flow at 400 °C for 10 min is necessary.

### 2.2. Catalysts Characterization

Table 1 summarizes the properties of the prepared catalysts. The designation of the samples shows the atomic Mn/Pd ratio and the treatment temperature, which is the final temperature of the cymantrene decomposition. As Table 1 shows, the Brunauer–Emmett–Teller (BET) surface area of the samples is the same within the margin of error, which indicates that the addition of Mn has no significant effect on the surface area of the catalysts. However, the samples show quite a different behavior in H$_2$ chemisorption.

Table 1. Physicochemical data of alumina supported catalysts.

| Sample | Catalyst Composition [1] | Hydrogen Treatment Temperature, °C | BET Surface Area, m²/g | $H_2$ Adsorption, µmol/g cat | Selectivity to Ethylene [2] at 40 °C, % |
|---|---|---|---|---|---|
| Pd-250 | 0.068%Pd | 250 | 133 | 1.20 | 70 |
| PdMn-1-250 | 0.068%Pd-0.029%Mn | 250 | 133 | 0.18 | 91 |
| PdMn-1-330 | 0.068%Pd-0.029%Mn | 330 | 131 | 1.10 | 80 |
| PdMn-2-300 | 0.068%Pd-0.063%Mn | 300 | 129 | 0.03 | 92 |
| PdMn-2-350 | 0.068%Pd-0.063%Mn | 350 | 128 | 0.04 | 89 |

[1] Hereafter, all catalyst compositions are in wt. %, [2] Acetylene conversion is 60%.

The non-promoted Pd/Al$_2$O$_3$ catalyst uptakes a significant amount of H$_2$ (1.2 µmol/g), but any addition of Mn decreases the H$_2$ adsorption. For example, an addition of 0.029% Mn (PdMn-1-250 and PdMn-1-330) decreases the H$_2$ adsorption to 1.1 µmol/g (for the sample treated at 330 °C) and to 0.18 µmol/g (for the sample treated at 250 °C). This trend is enhanced by a further addition of Mn: both PdMn-2-300 and PdMn-2-350 uptake significantly less H$_2$ (0.03 and 0.04 µmol/g). It should be noted that there are two possible reasons for the decreasing H$_2$ adsorption: the shielding of Pd with Mn atoms and the blocking of H$_2$ adsorption sites by residual CO ligands. Moreover, the selectivity of Pd-Mn catalysts to ethylene is correlated with their H$_2$ adsorption, as shown in Table 1.

As depicted in Figure 2, the non-modified Pd-250 desorbs H$_2$ in the range of 80–250 °C, indicating a desorption of weakly bonded hydrogen at low temperatures and strongly bonded hydrogen (or Pd hydride decomposition) at a temperature ramp.

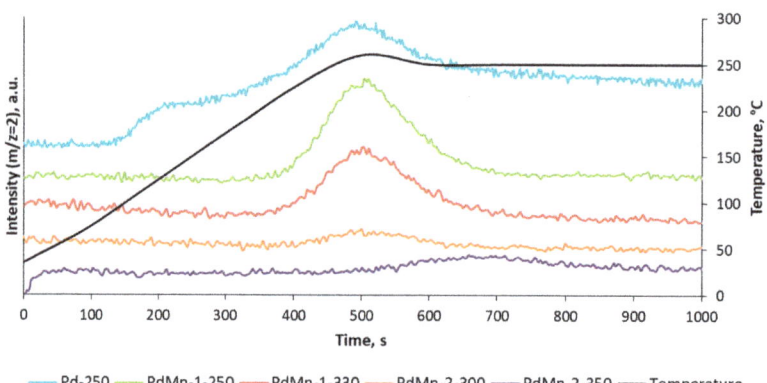

Figure 2. TPD-H$_2$ spectra of the samples.

The samples with a low Mn content (PdMn-1-250 and PdMn-1-330) do not desorb H$_2$ at temperatures below 230 °C, which indicates the presence of strongly bonded hydrogen (or Pd hydride). The samples (PdMn-2-300 and PdMn-2-350) with a high Mn content demonstrate only an insignificant H$_2$ desorption at a temperature of 250 °C, and these findings correlate with the chemisorption data (Table 1). Decreasing strongly chemisorbed hydrogen is recommended for acetylene selective hydrogenation as reported in [4,25,47].

As depicted on the TEM images of the PdMn-2-300, Pd nanoparticles (NP) with a lattice spacing of about 0.228 nm are found, which are indexed as the (111) plane for cubic palladium doped with clusters from single Mn atoms (Figure 3a) and Mn crystallites (Figure 3b) [48,49]. Depending on the lattice spacing, the fringes on the TEM images could be assigned to Pd nanoparticles or Mn crystallites and in some cases to manganese oxides with a lattice spacing of about 0.47–0.49 nm [49,50]. Due to the overlapping of Mn crystallites on Pd NPs, it is difficult to measure accurately the palladium nanoparticles' size and their distributions [51], but TEM images show Pd NPs in the range of 5–10 nm with a mean particle size of about 6.7 ± 0.2 nm (Figure 3c), in agreement with the literature data [48].

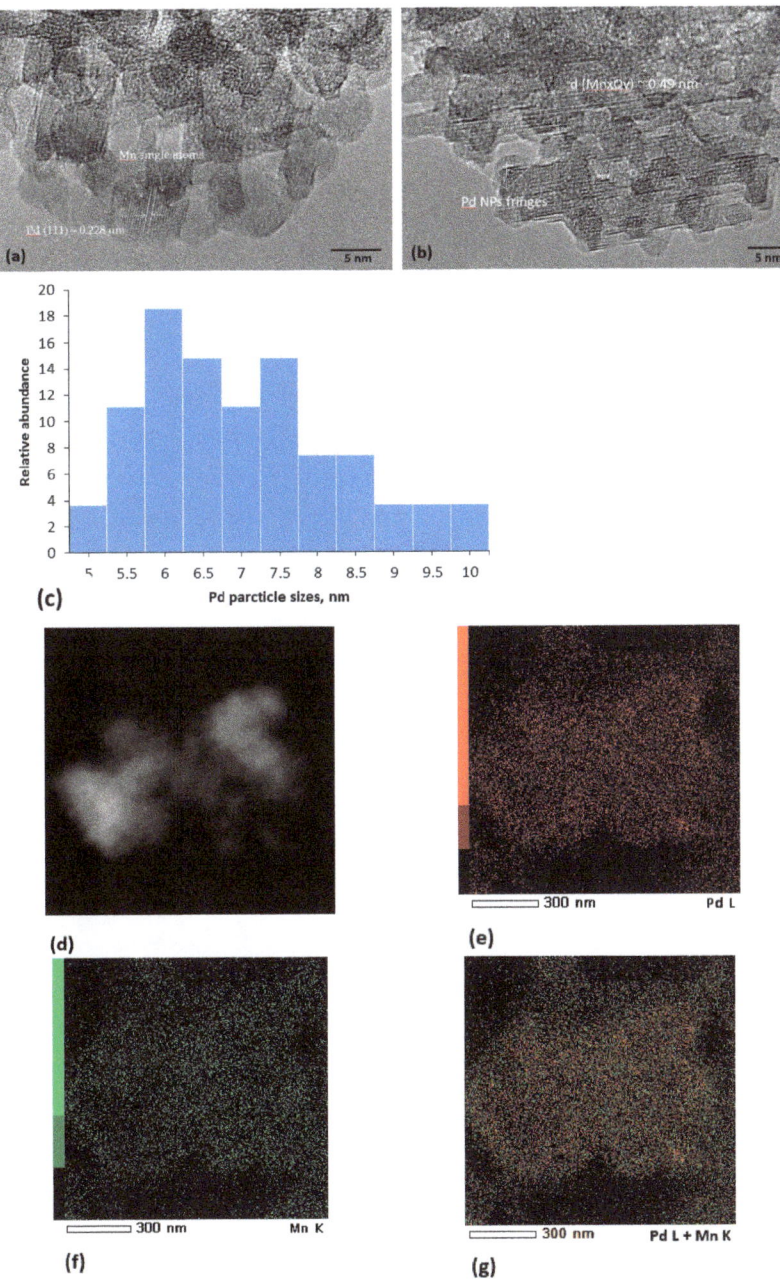

**Figure 3.** TEM imaging of the sample 4: (**a,b**) TEM image; (**c**) particle size distribution; (**d**) STEM image; (**e**) Pd mapping (L line); (**f**) Mn mapping (K line); (**g**) Pd (L) + Mn (K) mapping overlay.

Figure 3d–f show a STEM image and its EDX mapping of PdMn-2-300. It is clear that Pd and Mn are uniformly distributed over the alumina support with high dispersion. As shown in Figure 3g, both metals are in close contact.

XRD found no reflections, which could be related to Pd and Mn due to the low metal content, as Figure 4 shows.

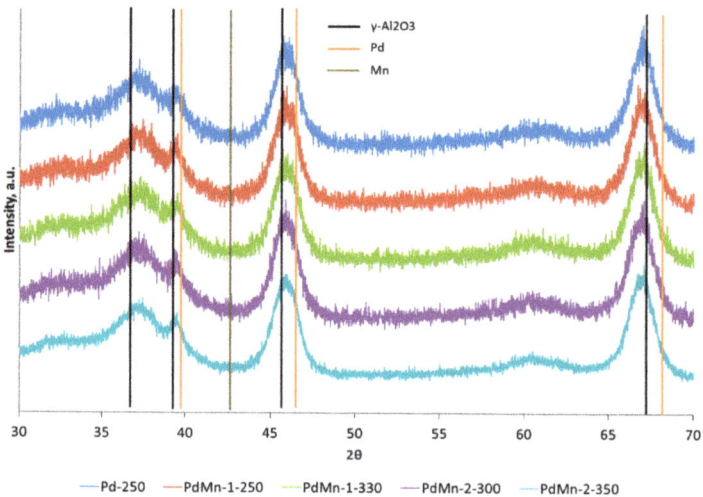

**Figure 4.** XRD patterns of Pd-Mn/Al$_2$O$_3$ samples.

Additional information about the chemisorption properties of the catalysts is obtained using DRIFT spectroscopy of PdMn-1-330. Figure 5 shows two spectra of the catalyst samples. For the first measurement, one sample is just treated in a vacuum for 2 h. For the second measurement, another sample is preliminarily treated with H$_2$ at 250 °C (30 min), acetylene at 20 °C (10 min), and H$_2$ at 20 °C (10 min) with a final purge with Ar at 20 °C (10 min) and treated in a vacuum for 2 h.

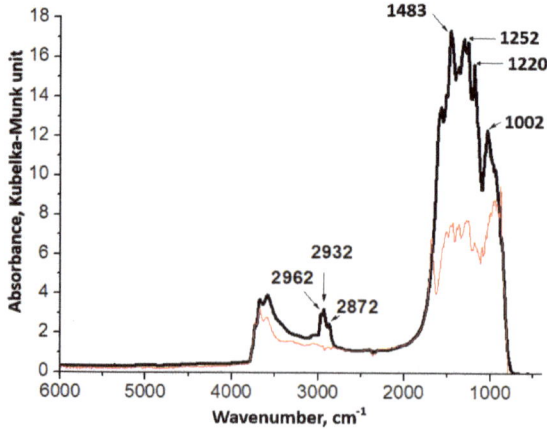

**Figure 5.** Diffuse reflectance infrared Fourier transform spectroscopy (DRIFTS) spectra of PdMn-1-330 without pretreatment (red line) and treated with C$_2$H$_2$ and H$_2$ (black line).

As the spectra show, there are two regions: 2500–3800 cm$^{-1}$, which corresponds to the vibrational spectra of O–H and C–H bonds, and 700–2400 cm$^{-1}$, ascribed to the vibrational spectra of Al$_2$O$_3$, adsorbed water, carbonyls, and others [52]. After the pretreatment, seven additional bands are observed: 2962, 2932, 2872, 1483, 1252, 1220, and 1002 cm$^{-1}$. The bands 2962, 2932, and 2872 cm$^{-1}$ may be ascribed to C–H stretching in the $C_2H_6$ molecule [53] and the bands 1220 and 1252 cm$^{-1}$ to vibrations of C–C bonds in the $C_2H_2$ molecule [54]. The band 1002 cm$^{-1}$ may be assigned to C=C bending in the $C_2H_4$ [53]. It should be stressed that all bands above are observed only after the treatment of PdMn-1-330 with $C_2H_2$. After vacuum treatment of the sample (2 h, 200 °C), the intensity of the spectra in the region of 2962–2872 cm$^{-1}$ decreases slightly, which points out the strong chemisorption of the species above.

The DRIFT spectra of adsorbed CO are considered in the Supplementary Materials.

As shown by the DRIFT, the CO adsorption over the Pd-Mn catalysts was weak and negligible (at most 0.015 units Kubelka-Munk). After the vacuum treatment at room temperature, all peaks in the range of 2195–1871 cm$^{-1}$ disappeared. So, one may conclude that there is an absence of strong CO chemisorption on the catalysts.

*2.3. Catalytic Tests*

Figure 6a shows the conversion of acetylene (X) as a function of the contact time (t) for the samples at 40 °C. For all of the Mn-promoted catalysts, the conversion values are at the same level (within the margin of error), regardless of the Mn addition and treating temperature. Moreover, at a given contact time, the Mn-promoted samples provide a significantly higher conversion (~20%) compared to that of Pd-250. The linear form of X(t) lines indicates the overall zero order by reagents.

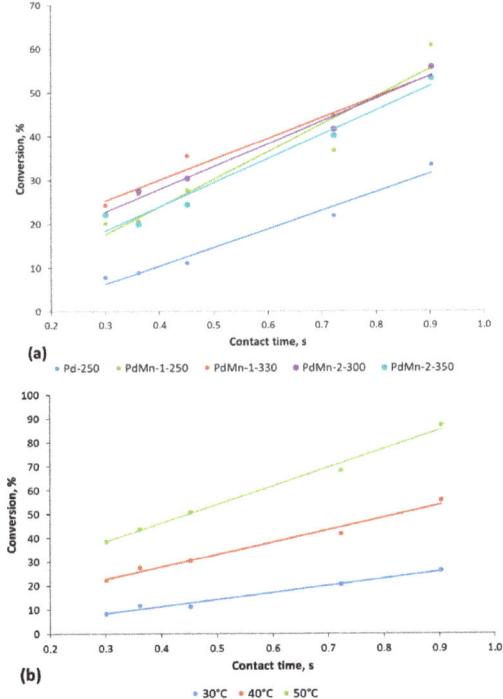

**Figure 6.** Acetylene conversion vs. contact time: (**a**) for the Pd-Mn/$Al_2O_3$ catalysts; (**b**) for PdMn-2-300 at 30, 40, and 50 °C.

Based on these experimental data and taking into account the selectivity obtained (Figure 7), we may consider a mass ratio of Mn/Pd ~ 1 (atomic ratio Mn/Pd ~ 2) and a treatment temperature of 300 °C for 30 min as optimal, corresponding to the PdMn-2-300 sample.

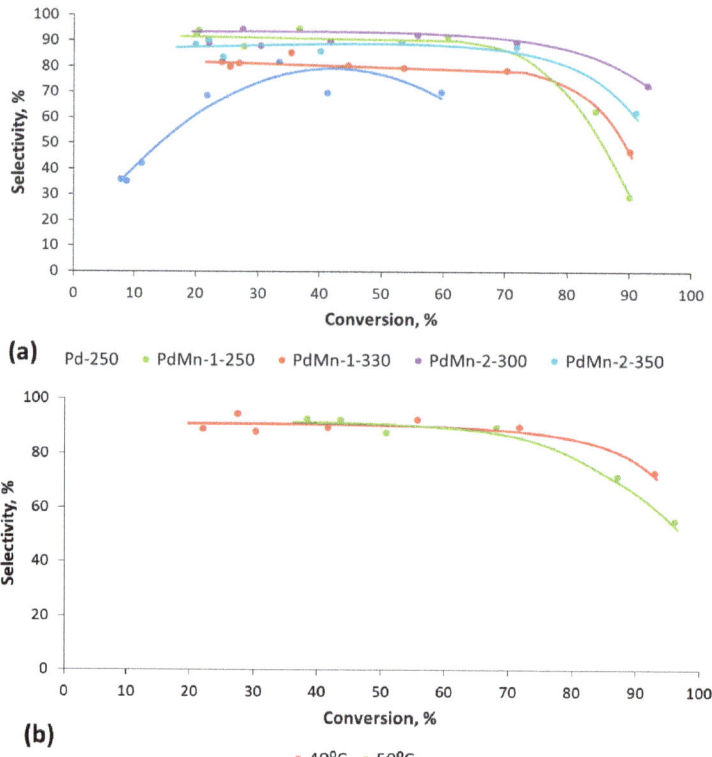

**Figure 7.** Ethylene selectivity vs. conversion at 40 °C for alumina-supported Pd-Mn catalysts: (**a**) effect of treatment temperature and Mn content; (**b**) PdMn-2-300 at 40 and 50 °C.

Figure 6b shows the conversion vs. contact time for PdMn-2-300 at 30, 40, and 50 °C. At all temperatures, the X(t) lines are straight, so the zero order by reagents is kept.

Figure 7 shows the selectivity to ethylene on the conversion for all samples at 40 °C. For Pd-250, the selectivity is the lowest and the curve has a maximum, which is typical in the case of a consecutive scheme of acetylene hydrogenation to ethylene and ethane:

$$C_2H_2 \rightarrow C_2H_4 \rightarrow C_2H_6 \qquad (1)$$

in accordance with [55].

However, Mn-containing samples maintain a selectivity at a level of 80%–92% up to acetylene conversions of more than 70%, which implies the parallel scheme of acetylene hydrogenation to ethylene and ethane:

$$C_2H_6 \leftarrow C_2H_2 \rightarrow C_2H_4 \qquad (2)$$

as previously observed on Pd-Fe/Al$_2$O$_3$ catalysts [22]. The PdMn-1-250 sample shows a selectivity of ~ 91% at a conversion of below 70%. Increasing the treating temperature to 330 °C (PdMn-1-330), and thus eliminating the strongly chemisorbed CO ligands, decreases the selectivity to ~80% at the same conversions.

Both samples with high Mn contents show a better selectivity at high conversions. The most selective is PdMn-2-300, of which the selectivity is ~92% in the conversion range of below 70%. The result is in accordance with the data published [56], where a catalyst having an Mn/Pd atomic ratio of two provides the highest selectivity to 1,3-butadiene in vinylacetylene hydrogenation.

Figure 7b shows the influence of the reaction temperature on the S(X) curve for PdMn-2-300 at 40 and 50 °C. It is obvious that the selectivity is the same (~90%) up to a conversion of ~70% irrespective of the temperature, which indicates that the activation energies of both reaction routes in scheme two are very close.

Our catalyst advantages are illustrated in Table 2, comprising the characteristics of the best Pd-containing catalysts in acetylene hydrogenation.

**Table 2.** Comparative characteristics of the best Pd-containing catalysts in acetylene hydrogenation.

| Article | Catalyst | T, K | P, bar | X, % | S, % | Activity, mol/g Pd/h | TOF (Turnover Frequency), $s^{-1}$ |
|---|---|---|---|---|---|---|---|
| [12] | $Pd_{30}Ga_{70}$ | 473 | 1 | 99 | 71 | 0.012 [1] | - |
| [24] | $Pd-Au/TiO_2$ | 343 | 1 | 100 | 45 | ~0.283 [2] | - |
| [25] | $Pd-In/Al_2O_3$ | 333 | 21 | ~85 | ~40 | - | 0.8 |
| [35] | Pd/Fiberglass | 328 | 1 | 80 | 60 | - | 0.55 |
| [8] | $Pd_{20}Ag_{80}/Al_2O_3$ | 303 | 10 | 67 | 72 | - | 0.5 |
| [22] | $Pd-Fe/Al_2O_3$ | 318 | 1 | 87 | 88 | 1.67 | 0.31 |
| This work | PdMn-2-300 | 313 | 1 | 87 | 81 | 4.22 | 0.74 |

[1] Recalculated from g/g cat/h, [2] Calculated by authors using data [24]: acetylene concentration and conversion, GHSV, Pd content and Pd density.

As Table 2 shows, the catalysts developed are of the same order of activity (in terms of turnover frequency (TOF)) but exceed the known one in ethylene yield (the product of X and S, calculated by acetylene (defficiency) conversion without hydrogen (excess)) and molar activity under mild conditions.

One may conclude, therefore, that modification with manganese improves both the activity (in terms of mol/$g_{Pd}$/h) and selectivity of palladium catalysts in acetylene hydrogenation.

For qualitative evaluation of the catalyst stability on a laboratory scale, the selectivity to the $C_4$ compound (namely, 1,3-butadiene as an initial dimerization product of the acetylenic species [9]) is used, which is a fundamental indicator of the palladium-containing catalyst stability in the selective hydrogenation of acetylene [8]. The absence of $C_4$ hydrocarbons is confirmed by GC and MS-analyses for all set experiments performed. More details about the analyses and the stability of the catalysts are shown in the Supplementary Materials. Based on the evidence above, we consider the Pd-Mn/$Al_2O_3$ catalysts to be stable for at least 5 h.

## 3. Materials and Methods

### 3.1. Chemical Reagents

Microspherical γ-$Al_2O_3$—SKTB Katalizator (Novosibirsk, Russia); cymantrene-cyclopentadienylmanganese tricarbonyl or $(CO)_3Mn$-(cyclo-$C_5H_5$) (Sigma-Aldrich, St. Louis, MO, USA); $PdCl_2$—Aurat (Moscow, Russia); Ar (99,993%), $H_2$ (99,99%), He (99,995%), $C_2H_2$ (99,1%), $C_2H_4$ (99,9%)—NII KM (Moscow, Russia); $NH_3 \cdot H_2O$ (~25%)—ECOS-1 (Moscow, Russia).

### 3.2. Catalysts Preparation

The initial catalyst 0.068% Pd/$Al_2O_3$ was prepared by a wet impregnation of γ-$Al_2O_3$ (preliminary calcined for 3 h at 600 °C) with an aqueous ammonia solution of $PdCl_2$ at pH = 12 (24 h). After a vacuum evaporation of the solvent, the catalyst was dried out at 70 °C for 12 h. Then, the catalyst was reduced with $H_2$ (20 mL/min) at 250 °C for 1 h.

Pd-Mn/$Al_2O_3$ samples were prepared by a wet impregnation of the reduced Pd/$Al_2O_3$ sample with a cymantrene solution in $n$-hexane. After the vacuum evaporation of the solvent, the samples were treated in an $H_2$ flow (20 mL/min) at 250–350 °C for 1 h. The effluent gas was analyzed with a quadrupole mass spectrometer QMS-200 (Stanford Research Systems, Sunnyvale, CA, USA).

### 3.3. Catalyst Characterization

The BET surface area was measured using Gemini VII (Micromeritics Instrument Corp.; Norcross, GA, USA). The samples were degassed at 150 °C for 3 h. The specific surface area was calculated using the BET model for adsorption data in the range of relative pressures $P/P_0$ = 0.05–0.30.

The metal content of the samples was measured by atomic absorption spectrometry (Perkin-Elmer-AAS, Waltham, MA, USA).

Pulse chemisorption of $H_2$ and TPD-$H_2$ was performed by AutoChem 2950HP (Micromeritics Instrument Corp.; Norcross, GA, USA). The samples were preliminarily reduced with $H_2$ at 250 °C for 1 h, purged with Ar for 30 min and cooled down to 35 °C. The pulse chemisorption was performed with a mixture 10% $H_2$ + Ar (balance), with a pulse volume of 0.5 mL, in an Ar flow (40 mL/min). TPD-$H_2$ was performed in an Ar flow (40 mL/min) at a heating rate of 30 K/min to 250 °C.

Transmission electron microscopy (TEM) analysis was carried out using a JEOL JEM-2100 microscope (Jeol Ltd.; Tokyo, Japan) with a 200 kV electron beam energy-dispersive X-ray analyzer (EDX). The mapping of the elements was carried out by scanning transmission electron microscopy (STEM). The samples were milled in an Eppendorf with a glass rod and ultrasonically suspended in isopropanol.

Phase analysis was performed using X-ray powder diffractometer BrukerD2 (Billerica, MA, USA), Cu K$\alpha$ ($\lambda$ = 1.5406 Å), 2$\theta$ values varied from 5° to 80°.

Diffuse reflectance infrared Fourier transform spectroscopy was done using a NICOLET Protégé 460 (Nicolet, Madison, WI, USA) in the range of 6000–400 cm$^{-1}$ at a resolution of 4 cm$^{-1}$. For each sample, 500 spectra were recorded to get a good signal-noise ratio. $CaF_2$ was used as a standard. The spectra were processed with OMNIC software.

### 3.4. Catalystic Tests

Acetylene hydrogenation was performed in a quartz reactor at atmospheric pressure using AutoChem 2950HP (Micromeritics Instrument Corp.; Norcross, GA, USA). At a given temperature, the flow rate of the reaction mixture was changed to get various values of conversion and selectivity. The contact time was in the range of 0.26–1.81 s$^{-1}$, the reaction temperature was in the range of 30–50 °C.

A mixture of 1.94%$H_2$ + 1.05%$C_2H_2$ + 5.01%$C_2H_4$ + Ar (balance) was used as a modeling feed preliminarily prepared in a cylinder. The effluent gas was analyzed online using a quadrupole mass spectrometer QMS-200 (Stanford Research Systems, Sunnyvale, CA, USA) and off-line using FID and TCD detectors in a GC experimental laboratory chromatograph (Gubkin University—Chromos, on the basis of GC-1000 model, Moscow—Dzerjinsk, Russia) using a packed column with HyeSep N. At given operating conditions (temperature, flow rate), the effluent gas was analyzed three times and the final concentration was calculated as the mean value of the three analyses. The carbon balance was closed within 4%.

The acetylene conversion was calculated by the equation:

$$X = \frac{C_{C2H2}^{in} - C_{C2H2}^{out}}{C_{C2H2}^{in}} \times 100\% \tag{3}$$

and ethylene selectivity by:

$$S_{C2H4} = \frac{C_{C2H4}^{out} - C_{C2H4}^{in}}{C_{C2H2}^{in} - C_{C2H2}^{out}} \times 100\% \tag{4}$$

## 4. Conclusions

A number of Pd-Mn/Al$_2$O$_3$ catalysts were designed by the decomposition of cymantrene on reduced Pd/Al$_2$O$_3$ in an H$_2$ atmosphere. The formation of bimetallic catalysts was studied by mass spectrometry analysis of the decomposition products. It was found that the decomposition of cymantrene takes place with hydrogenation of cyclopentadienyl ligands to cyclopentene and cyclopentane, and CO ligands are partially removed by conversion to methane. The catalysts are characterized using N$_2$ adsorption, H$_2$ pulse chemisorption, TPD-H$_2$, TEM, EDX, XRD, and DRIFT spectroscopy. Using the organic precursor—cymantrene provides a high and uniform distribution of Mn over Pd. The addition of manganese changes the H$_2$ chemisorption and desorption properties of the catalyst: the Pd-Mn/Al$_2$O$_3$ samples have shown either a strong chemisorption of H$_2$ or an insignificant H$_2$ chemisorption. At the same time, unsaturated C$_2$ hydrocarbons are strongly chemisorbed on Pd-Mn/Al$_2$O$_3$ samples and cannot be removed even under vacuum treatment at elevated temperature. Catalytic tests of the novel Pd-Mn/Al$_2$O$_3$ catalysts in hydrogenation of acetylene have shown a higher activity and selectivity thereof to ethylene (up to 20% higher) compared to the non-promoted Pd/Al$_2$O$_3$ catalyst. The optimal Mn/Pd ratio and treatment temperature are found. The overall reaction order by reagents is zero for all catalysts, but modification with Mn changes the reaction route from a consecutive pathway for Pd/Al$_2$O$_3$ to a parallel one for Pd-Mn/Al$_2$O$_3$ catalysts.

**Supplementary Materials:** The following are available online at http://www.mdpi.com/2073-4344/10/6/624/s1, Figure S1: Peaks of m/z in decomposition of cymantrene in H$_2$.

**Author Contributions:** Conceptualization, V.S., D.M.; methodology, D.M., E.S.; software, V.L., E.I., M.K.; validation, A.G., V.S.; formal analysis, V.V., D.M.; resources, V.V., A.G.; data curation, D.M., V.S., E.S., M.K.; writing—original draft preparation, D.M., V.S.; writing—review and editing, D.M., V.S., A.G.; visualization, D.M., V.L.; supervision, V.V., V.S.; project administration, V.V.; funding acquisition, A.G., E.I. All authors have read and agreed to the published version of the manuscript.

**Funding:** This work was financially supported by the Ministry of Science and Higher Education of the Russian Federation in the part of the analysis technique development in Gubkin University (experimental laboratory gas chromatograph, analysis of hydrocarbons and hydrogen, agreement number 075-11-2019-037 (agreement number between Gubkin University and LLC "Chromos Engineering" 555-19) and as a part of the state task of Gubkin University (synthesis of catalysts, catalytic and phys-chem experiments), project number FSZE-2020-0007 (0768-2020-0007, A.G., E.I., V.V.).

**Acknowledgments:** The authors thank Olga P. Tkachenko (N.D. Zelinsky Institute of Organic Chemistry) for DRIFT spectra measurements and interpretation. A.G., E.I., V.S., D.M. and V.V. thank LLC "Chromos Engineering" and Andrei Pakhomov.

**Conflicts of Interest:** The authors declare no conflict of interest.

## References

1. Arnold, H.; Döbert, F.; Gaube, J. Hydrogenation reactions. In *Handbook of Heterogeneous Catalysis*, 2nd ed.; Ertl, G., Knözinger, H., Schüth, F., Weitkamp, J., Eds.; Wiley-VCH Verlag GmbH&Co. KGaA: Weinheim, Germany, 2008; Volume 7, pp. 3266–3359.
2. Stytsenko, V.D.; Mel'nikov, D.P. Selective Hydrogenation of Diene and Acetylene Compounds on Metal-Containing Catalysts. *Russ. J. Phys. Chem. A* **2016**, *90*, 932–942. [CrossRef]
3. Huang, D.C.; Chang, K.H.; Pong, W.F.; Tseng, P.K.; Hung, K.J.; Huang, W.F. Effect of Ag-promotion on Pd catalysts by XANES. *Catal. Lett.* **1998**, *53*, 155–159. [CrossRef]
4. Zhang, Q.; Li, J.; Liu, X.; Zhu, Q. Synergetic effect of Pd and Ag dispersed on Al$_2$O$_3$ in the selective hydrogenation of acetylene. *Appl. Catal. A Gen.* **2000**, *197*, 221–228. [CrossRef]
5. Pei, G.X.; Liu, X.Y.; Wang, A.; Lee, A.F.; Isaacs, M.A.; Li, L.; Pan, X.; Yang, X.; Wang, X.; Tai, Z.; et al. Ag alloyed Pd single-atom catalysts for efficient selective hydrogenation of acetylene to ethylene in excess ethylene. *Acs Catal.* **2015**, *5*, 3717–3725. [CrossRef]
6. Zhang, Y.; Diao, W.; Williams, C.T.; Monnier, J.R. Selective hydrogenation of acetylene in excess ethylene using Ag- and Au-Pd/SiO$_2$ bimetallic catalysts prepared by electroless deposition. *Appl. Catal. A Gen.* **2014**, *469*, 419–426. [CrossRef]

7. Zhang, Y.; Diao, W.; Monnier, J.R.; Williams, C.T. Pd-Ag/SiO$_2$ bimetallic catalysts prepared by galvanic displacement for selective hydrogenation of acetylene in excess ethylene. *Catal. Sci. Technol.* **2015**, *5*, 4123–4132. [CrossRef]
8. Kuhn, M.; Lucas, M.; Claus, P. Long-time stability vs deactivation of Pd-Ag/Al$_2$O$_3$ egg-shell catalysts in selective hydrogenation of acetylene. *Ind. Eng. Chem. Res.* **2015**, *54*, 6683–6691. [CrossRef]
9. Ahn, I.Y.; Lee, J.H.; Kim, S.K.; Moon, S.H. Three-stage deactivation of Pd/SiO$_2$ and Pd-Ag/SiO$_2$. *Appl. Catal. A Gen.* **2009**, *360*, 38–42. [CrossRef]
10. Shin, E.W.; Choi, C.H.; Chang, K.S.; Na, Y.H.; Moon, S.H. Properties of Si-modified Pd catalyst for selective hydrogenation of acetylene. *Catal. Today* **1998**, *44*, 137–143. [CrossRef]
11. Osswald, J.; Giedigkeit, R.; Jentoft, R.E.; Armbrüster, M.; Girgsdies, F.; Kovnir, K.; Ressler, T.; Grin, Y.; Schlögl, R. Palladium–gallium intermetallic compounds for the selective hydrogenation of acetylene: Part I: Preparation and structural investigation under reaction conditions. *J. Catal.* **2008**, *258*, 210–218. [CrossRef]
12. Osswald, J.; Kovnir, K.; Armbrüster, M.; Giedigkeit, R.; Jentoft, R.E.; Wild, U.; Grin, Y.; Schlögl, R. Palladium–gallium intermetallic compounds for the selective hydrogenation of acetylene: Part II: Surface characterization and catalytic performance. *J. Catal.* **2008**, *258*, 219–227. [CrossRef]
13. Armbrüster, M.; Kovnir, K.; Behrens, M.; Teschner, D.; Grin, Y.; Schlögl, R. Pd-Ga intermetallic compounds as highly selective semihydrogenation catalysts. *J. Am. Chem. Soc.* **2010**, *132*, 14745–14747. [CrossRef] [PubMed]
14. Esmaeili, E.; Rashidi, A.M.; Khodadadi, A.A.; Mortazavi, Y.; Rashidzadeh, M. Palladium-Tin nanocatalyst in high concentration acetylene hydrogenation: A novel deactivation mechanism. *Fuel Process. Technol.* **2014**, *120*, 112–122. [CrossRef]
15. Kim, S.K.; Lee, J.H.; Ahn, I.Y.; Kim, W.J.; Moon, S.H. Performance of Cu-promoted Pd catalysts prepared by adding Cu using a surface redox method in acetylene hydrogenation. *Appl. Catal. A Gen.* **2011**, *401*, 12–19. [CrossRef]
16. Cao, X.; Mirjalili, A.; Wheeler, J.; Xie, W.; Jang, B.W.L. Investigation of the preparation methodologies of Pd-Cu single atom alloy catalysts for selective hydrogenation of acetylene. *Front. Chem. Sci. Eng.* **2015**, *9*, 442–449. [CrossRef]
17. Studt, F.; Abild-Pedersen, F.; Bligaard, T.; Sørensen, R.Z.; Christensen, C.H.; Nørskov, J.K. Identification of non-precious metal alloy catalysts for selective hydrogenation of acetylene. *Science* **2008**, *320*, 1320–1322. [CrossRef]
18. McCue, A.J.; Guerrero-Ruiz, A.; Ramirez-Barria, C.; Rodríguez-Ramos, I.; Anderson, J.A. Selective hydrogenation of mixed alkyne/alkene streams at elevated pressure over a palladium sulfide catalyst. *J. Catal.* **2017**, *355*, 40–52. [CrossRef]
19. McCue, A.J.; McRitchi, C.J.; Shepherd, A.M.; Anderson, J.A. Cu/Al$_2$O$_3$ catalysts modified with Pd for selective acetylene hydrogenation. *J. Catal.* **2014**, *319*, 127–135. [CrossRef]
20. Chinayon, S.; Mekasuwandumrong, O.; Praserthdam, P.; Panpranot, J. Selective hydrogenation of acetylene over Pd catalysts supported on nanocrystalline α-Al$_2$O$_3$ and Zn-modified α-Al$_2$O$_3$. *Catal. Commun.* **2008**, *9*, 2297–2302. [CrossRef]
21. Mashkovsky, I.S.; Baeva, G.N.; Stakheev, A.Y.; Vargaftik, M.N.; Kozitsyna, N.Y.; Moiseev, I.I. Novel Pd-Zn/C catalyst for selective alkyne hydrogenation: Evidence for the formation of Pd-Zn bimetallic alloy particles. *Mendeleev Commun.* **2014**, *24*, 355–357. [CrossRef]
22. Stytsenko, V.D.; Mel'nikov, D.P.; Tkachenko, O.P.; Savel'eva, E.V.; Semenov, A.P.; Kustov, L.M. Selective Hydrogenation of Acetylene and Physicochemical Properties of Pd-Fe/Al$_2$O$_3$ Bimetallic Catalysts. *Russ. J. Phys. Chem. A* **2018**, *92*, 862–869. [CrossRef]
23. Jia, J.; Haraki, K.; Kondo, J.N.; Domen, K.; Tamaru, K. Selective hydrogenation of acetylene over Au/Al$_2$O$_3$ catalyst. *J. Phys. Chem. B* **2000**, *104*, 11153–11156. [CrossRef]
24. Choudhary, T.V.; Sivadinarayana, C.; Datye, A.K.; Kumar, D.; Goodman, D.W. Acetylene hydrogenation on Au-based catalysts. *Catal. Lett.* **2003**, *86*, 1–8. [CrossRef]
25. Cao, Y.; Sui, Z.; Zhu, Y.; Zhou, X.; Chen, D. Selective Hydrogenation of Acetylene over Pd-In/Al$_2$O$_3$ Catalyst: Promotional Effect of Indium and Composition-Dependent Performance. *ACS Catal.* **2017**, *7*, 7835–7846. [CrossRef]

26. Spanjers, C.S.; Held, J.T.; Jones, M.J.; Stanley, D.D.; Sim, R.S.; Janik, M.J.; Rioux, R.M. Zinc inclusion to heterogeneous nickel catalysts reduces oligomerization during the semi-hydrogenation of acetylene. *J. Catal.* **2014**, *316*, 164–173. [CrossRef]
27. Tkachenko, O.P.; Stakheev, A.Y.; Kustov, L.M.; Mashkovsky, I.V.; van den Berg, M.; Grünert, W.; Kozitsyna, N.Y.; Dobrokhotova, Z.V.; Zhilov, V.I.; Nefedov, S.E.; et al. An easy way to Pd-Zn nanoalloy with defined composition from a heterometallic Pd(μ-OOCMe)$_4$Zn(OH$_2$) complex as evidenced by XAFS and XRD. *Catal. Lett.* **2006**, *112*, 155–161. [CrossRef]
28. Vinokurov, V.A.; Stavitskaya, A.V.; Chudakov, Y.A.; Glotov, A.P.; Ivanov, E.V.; Gushchin, P.A.; Lvov, Y.M.; Maximov, A.L.; Muradov, A.V.; Karakhanov, E.A. Core-shell nanoarchitecture; Schiff-base assisted synthesis of ruthenium in clay nanotubes. *Pure Appl. Chem.* **2018**, *90*, 825–832. [CrossRef]
29. Vinokurov, V.A.; Stavitskaya, A.V.; Glotov, A.P.; Novikov, A.A.; Zolotukhina, A.V.; Kotelev, M.S.; Gushchin, P.A.; Ivanov, E.V.; Darrat, Y.; Lvov, Y.M. Nanoparticles Formed onto/into Halloysite Clay Tubules: Architectural Synthesis and Applications. *Chem. Rec.* **2018**, *18*, 858–867. [CrossRef]
30. Vinokurov, V.; Glotov, A.; Chudakov, Y.; Stavitskaya, A.; Ivanov, E.; Gushchin, P.; Zolotukhina, A.; Maximov, A.; Karakhanov, E.; Lvov, Y. Core/Shell Ruthenium–Halloysite Nanocatalysts for Hydrogenation of Phenol. *Ind. Eng. Chem. Res.* **2017**, *56*, 14043–14052. [CrossRef]
31. Glotov, A.; Levshakov, N.; Stavitskaya, A.; Artemova, M.; Gushchin, P.; Ivanov, E.; Vinokurov, V.; Lvov, Y. Templated Self-Assembly of Ordered Mesoporous Silica on Clay Nanotubes. *Chem. Commun.* **2019**, *55*, 5507–5510. [CrossRef]
32. Kang, J.H.; Shin, E.W.; Kim, W.J.; Park, J.D.; Moon, S.H. Selective hydrogenation of acetylene on Pd/SiO$_2$ catalysts promoted with Ti, Nb and Ce oxides. *Catal. Today* **2000**, *63*, 183–188. [CrossRef]
33. Kang, J.H.; Shin, E.W.; Kim, W.J.; Park, J.D.; Moon, S.H. Selective hydrogenation of acetylene on TiO$_2$-added Pd catalysts. *J. Catal.* **2002**, *208*, 310–320. [CrossRef]
34. Gulyaeva, Y.K.; Kaichev, V.V.; Zaikovskii, V.I.; Kovalyov, E.V.; Suknev, A.P.; Bal'zhinimaev, B.S. Selective hydrogenation of acetylene over novel Pd/fiberglass catalysts. *Catal. Today* **2015**, *245*, 139–146. [CrossRef]
35. Gulyaeva, Y.K.; Kaichev, V.V.; Zaikovskii, V.I.; Suknev, A.P.; Bal'zhinimaev, B.S. Selective hydrogenation of acetylene over Pd/Fiberglass catalysts: Kinetic and isotopic studies. *Appl. Catal. A Gen.* **2015**, *506*, 197–205. [CrossRef]
36. Shesterkina, A.A.; Kozlova, L.M.; Kirichenko, O.A.; Kapustin, G.I.; Mishin, I.V.; Kustov, L.M. Influence of the thermal treatment conditions and composition of bimetallic catalysts Fe-Pd/SiO$_2$ on the catalytic properties in phenylacetylene hydrogenation. *Russ. Chem. Bull.* **2016**, *65*, 432–439. [CrossRef]
37. Shesterkina, A.A.; Kozlova, L.M.; Mishin, I.V.; Tkachenko, O.P.; Kapustin, G.I.; Zakharov, V.P.; Vlaskin, M.S.; Zhuk, A.Z.; Kirichenko, O.A.; Kustov, L.M. Novel Fe-Pd/γ-Al2O3 catalysts for the selective hydrogenation of C≡C bonds under mild conditions. *Mendeleev Commun.* **2019**, *29*, 339–342. [CrossRef]
38. Rozovskii, A.Y.; Stytsenko, V.D.; Nizova, S.A.; Belov, P.S.; Dyakonov, A.Y. Catalyst for dehydrogenation of oxygen containing derivatives of the cyclohexane series into corresponding cyclic ketones and/or phenols. U.S. Patent 4,363,750, 14 December 1982.
39. Rozovskii, A.Y.; Stytsenko, V.D.; Nizova, S.A.; Belov, P.S.; Dyakonov, A.Y. Catalyst for dehydrogenation oxygen-containing derivatives of the cyclohexane series into the corresponding cyclic ketones and/or phenols. U.S. Patent 4,415,477, 15 November 1983.
40. Rozovskii, A.Y.; Stytsenko, V.D.; Nizova, S.A.; Belov, P.S.; Dyakonov, A.Y. Catalyst and process for dehydrogenation of oxygen-containing derivatives of the cyclohexane series into corresponding cyclic ketones and/orphenols. U.S. Patent 4,417,076, 22 November 1983.
41. Stytsenko, V.D. Surface modified bimetallic catalysts: Preparation, characterization, and applications. *Appl. Catal. A* **1995**, *126*, 1–26. [CrossRef]
42. Agnelli, M.; Louessard, P.; El Mansour, A.; Candy, J.P.; Bournonville, J.P.; Basset, J.M. Surface organometallic chemistry on metals preparation of new selective bimetallic catalysts by reaction of tetra-n-butyl tin with silica supported Rh, Ru and Ni. *Catal. Today* **1989**, *6*, 63–72. [CrossRef]
43. Adúriz, H.R.; Gígola, C.E.; Sica, A.M.; Volpe, M.A.; Touroude, R. Preparation and characterization of Pd-Pb catalysts for selective hydrogenation. *Catal. Today* **1992**, *15*, 459–467. [CrossRef]
44. Glotov, A.; Stytsenko, V.; Artemova, M.; Kotelev, M.; Ivanov, E.; Gushchin, P.; Vinokurov, V. Hydroconversion of Aromatic Hydrocarbons over Bimetallic Catalysts. *Catalysts* **2019**, *9*, 384. [CrossRef]
45. NIST Chemistry WebBook. Available online: https://webbook.nist.gov (accessed on 10 April 2020).

46. Wang, S.Y.; Moon, S.H.; Vannice, M.A. The Effect of SMSI (Strong Metal-Support Interaction) Behavior on CO Adsorption and Hydrogenation on Pd Catalysts. *J. Catal.* **1981**, *71*, 167–174. [CrossRef]
47. Lee, J.H.; Kim, S.K.; Ahn, I.Y.; Kim, W.J.; Moon, S.H. Performance of Ni-added Pd-Ag/Al$_2$O$_3$ catalysts in the selective hydrogenation of acetylene. *Korean J. Chem. Eng.* **2012**, *29*, 169–172. [CrossRef]
48. Li, Z.Y.; Liu, Z.L.; Liang, J.C.; Xu, C.W.; Lu, X. Facile synthesis of Pd-Mn$_3$O$_4$/C as high efficient electrocatalyst for oxygen evolution reaction. *J. Mater. Chem. A* **2014**, *2*, 18236–18240. [CrossRef]
49. Romano, C.A.; Zhou, M.; Song, Y.; Wysocki, V.H.; Dohnalkova, A.C.; Kovarik, L.; Paša-Tolić, L.; Tebo, B.M. Biogenic manganese oxide nanoparticle formation by a multimeric multicopper oxidase Mnx. *Nat. Commun.* **2017**, *8*, 1–8. [CrossRef] [PubMed]
50. Yasmin, S.; Cho, S.; Jeon, S. Electrochemically reduced grapheme-oxide supported bimetallic nanoparticles highly efficient for oxygen reduction reaction with excellent methanol tolerance. *Appl. Surf. Sci.* **2018**, *434*, 905–912. [CrossRef]
51. Wang, W.; Yuan, F.; Niu, X.; Zhu, Y. Preparation of Pd supported on La(Sr)-Mn-O Perovskite by microwave Irradiation Method and Its Catalytic Performances for the Methane Combustion. *Sci. Rep.* **2016**, *6*, 19511. [CrossRef]
52. Kustov, L.M. New trends in IR-spectroscopic characterization of acid and basic sites in zeolites and oxide catalysts. *Top. Catal.* **1997**, *4*, 131–144. [CrossRef]
53. Merck Webpage. Available online: https://www.sigmaaldrich.com/technical-documents/articles/biology/ir-spectrum-table.html (accessed on 18 May 2020).
54. Lapinski, M.P.; Ekerdt, J.G. Infrared Identification of Adsorbed Surface Species on Ni/SiO$_2$ and Ni/Al$_2$O$_3$ form Ethylene and Acetylene Adsorption. *J. Phys. Chem.* **1990**, *94*, 4599–4610. [CrossRef]
55. Urmès, C.; Schweitzer, J.M.; Cabiac, C.; Schuurman, Y. Kinetic Study of the Selective Hydrogenation of Acetylene over Supported Palladium under Tail-End Conditions. *Catalysts* **2019**, *9*, 180. [CrossRef]
56. Insorn, P.; Kitiyanan, B. Selective hydrogenation of mixed C$_4$ containing high vinyl acetylene by Mn-Pd, Ni-Pd and Ag-Pd on Al$_2$O$_3$ catalysts. *Catal. Today* **2015**, *256*, 223–230. [CrossRef]

© 2020 by the authors. Licensee MDPI, Basel, Switzerland. This article is an open access article distributed under the terms and conditions of the Creative Commons Attribution (CC BY) license (http://creativecommons.org/licenses/by/4.0/).

Article

# In Situ Generated Nanosized Sulfide Ni-W Catalysts Based on Zeolite for the Hydrocracking of the Pyrolysis Fuel Oil into the BTX Fraction

Tatiana Kuchinskaya *, Mariia Kniazeva, Vadim Samoilov and Anton Maximov

A.V. Topchiev Institute of Petrochemical Synthesis, Russian Academy of Sciences (TIPS RAS), 119991 Moscow, Russia; knyazeva@ips.ac.ru (M.K.); samoilov@ips.ac.ru (V.S.); max@ips.ac.ru (A.M.)
* Correspondence: kuchinskaya@ips.ac.ru; Tel.: +7-495-647-5927 (ext. 349)

Received: 9 September 2020; Accepted: 1 October 2020; Published: 7 October 2020

**Abstract:** The hydrocracking reaction of a pyrolysis fuel oil fraction using in situ generated nano-sized NiWS-sulfide catalysts is studied. The obtained catalysts were defined using X-ray photoelectron spectroscopy (XPS) and transmission electron microscopy (TEM). The features of catalytically active phase generation, as well as its structure and morphology were considered. The catalytic reactivity of in situ generated catalysts was evaluated using the hydrocracking reaction of pyrolysis fuel oil to obtain a light fraction to be used as a feedstock for benzene, toluene, and xylene (BTX) production. It was demonstrated that the temperature of 380 °C, pressure of 5 MPa, and catalyst-to-feedstock ratio of 4% provide for a target fraction (IPB −180 °C) yield of 44 wt %, and the BTX yield of reaching 15 wt %.

**Keywords:** BTX; dispersed catalysts; hydroconversion; pyrolysis fuel oil; zeolites; in situ formed catalysts; nanocatalyst

---

## 1. Introduction

The increasing development of environmentally friendly energy sources and the search for new ones such as the sun, hydrogen, biomass [1] has had a negative impact on the price and global demand for oil and natural gas used for fuel production. This has become a challenge for the oil producing industry since a large group of chemicals can still be obtained from oil and natural gas only. Thus, despite of the intensified studies on the possibility to obtain the benzene, toluene, and xylene (BTX) fraction from alternative sources such as plastic waste as well as bio oil components [2–6], oil resources are still the primary method of BTX fraction recovery [7–9]. The demand for aromatic hydrocarbons and BTX fraction as their primary source has been increasing from year to year because of the consistently growing polymers market. The main sources of the BTX fraction today are reforming and the process is mainly known as the "steam cracking." Benzene, toluene, and xylene (BTX) are widely used as the feedstock for the industrial production of chemicals. In its turn, heavy oil fractions rich in polyaromatic compounds (that originate also from the both catalytic reforming and the naphtha steam cracking) can be used for the BTX production [10,11].

Because of the increasing demand for low molecular weight olefins, the number of steam cracking facilities has increased as well. The annual average growth of ethylene and propylene consumption exceeds 4%. Steam cracking capacity will continue to grow driven by ethylene/propylene demand. This process results in a large number of aromatic compounds and low-value waste in the form of pyrolysis fuel oil (PFO). PFO is used as a component of boiler fuel oil, or for the production of petroleum resins. PFO is not currently being recycled properly. Residual fractions of the pyrolysis process, in particular pyrolysis fuel oil (PFO), can be used for the BTX production. As PFO consists mainly of diaromatic compounds such as naphthalene (35.98%), its ethyl-substituted equivalents of naphthalene (10.95%) and biphenyl (27.97%), it could be considered as a promising feedstock for the hydrocracking process that yields valuable products, such as gasoline or aromatics [12,13]. The hydrocracking of heavy oil residues is the wide-spread method used to lighten the composition of heavy oil residues and obtain a more valuable feedstock [14–17].

Supported metal catalysts are generally used as hydrocracking catalysts. Molybdenum or tungsten is used as an active metal. The catalysts are active in their sulfide form, with nickel or cobalt used as promoting substances. Acid materials such as zeolites, other aluminosilicates, or aluminum oxide are used as support [18–22]. Some studies describe the influence of oxide substrates on hydrogenation, hydrodesulfuration, and hydrocracking, stating that aluminum oxide-supported systems have a higher reactivity during hydrogenation [23–25]. Publications describe successful examples of the recovery of a commercially valuable BTX fraction from light catalytic gas oil (LCGO) on supported NiMo/Al2O3 catalysts, with a BTX fraction yield of 44 to 70 wt % [26]. In this case, the average yield of the BTX fraction varies from 15% to 50% when obtained through pyrolysis [2,6], and from 20% to 50% when obtained through hydrocracking [9,27].

Unsupported sulfide catalysts have been gaining popularity recently during studies of hydroprocessing in oil refining. This is due to their high catalytic reactivity as compared to supported hydrotreating catalysts [28,29], as well as their ability to offer a solution for the active center's deactivation problem caused by the deposition of coke and condensation products [30,31].

Unsupported sulfide catalysts on the basis of transition metals can be obtained using different methods: coacervation [32], codeposition [33], or precursor decomposition [34–36]. Precursors can include water-soluble [34,35] and oil-soluble [36] transition metal compounds. Despite the fact that the use of water-soluble compounds is more cost effective because of their low cost, they entail the formation of somewhat larger particles, which reduces the reactivity of the obtained catalyst [37]. Dispersed Ni (Co) –W (Mo) –S sulfide catalysts synthesized through in situ decomposition of various precursors are stable and characterized by high dispersiveness and a highly developed external surface [38].

Comparison of the reactivity of $MoS_2$ and $WS_2$ sulfide catalysts generated in situ from oil-soluble precursors during the hydrocracking of heavy residues has demonstrated the advantage of tungsten catalysts over molybdenum ones. $WS_2$ was more dispersive and demonstrated a higher hydrogenating reactivity as compared to the molybdenum equivalent [39].

As was established previously, the Ni–W–S catalyst exhibits reactivity in the course of coking resin hydrogenation. The optimal temperature of the pyrolysis oil hydrotreatment is 380 °C. It provides for the minimum amount of bicyclic and polycyclic aromatic hydrocarbons, and ensures no thermal cracking process. [40].

Previously, the hydrocracking of hydrocarbon fractions in the presence of the in situ generated catalysts was described in the works [41–45]. However, to the best of our knowledge, there are no precedents for the use of a bifunctional nanoheterogeneous catalytic system to obtain the BTX fraction from the pyrolysis fuel oil.

In continuation of work on the study of the regularities of hydrogenation of aromatized hydrocarbon fractions in the presence of nanosized in situ formed sulfide catalysts [35,46].

We offer a fundamentally new design opportunity for a quasi-homogeneous bifunctional catalytic system that can solve the problem of obtaining BTX from a pyrolysis fuel oil.

This study demonstrates the possibility of the use of the unsupported NiWS-HY catalysts generated in situ in the hydrocarbon feedstock from such oil-soluble precursors as tungsten hexacarbonyl ($W(CO)_6$) and Nickel(II)2-ethylhexanoate ($Ni(C_7H_{15}COO)_2$) with HY zeolite addition for the selective recovery of the BTX fraction from pyrolysis fuel oil.

## 2. Results

### 2.1. Catalytic Properties

2.1.1. Naphthalene Hydrocracking with NiWS-HY Catalysts

The reactivity of in situ generated catalysts as well as the possibility of the BTX fraction selective recovery were studied by the example of a model feedstock, for which purpose 10% solution of naphthalene in hexadecane was chosen as the feed mixture. The study was carried out in an autoclave-type batch reactor at P = 5 MPa, t = 5 h, and W:Ni ratio = 1:2. With a view to identifying the influence of temperature on product composition and conversion, experiments on in situ generated NiWS-HY catalyst were conducted at 350 °C, 380 °C, and 400 °C by varying the zeolite content at a level of 0.5 wt % and 1 wt %, and tungsten content at a level of 1 wt %, 2 wt %, and 4 wt %. Blank test showed that hexadecane does not react in these conditions.

The conversion of naphthalene under the selected conditions (Figure 1) was high and reached 84%–93%. Temperature increase up to 400 °C resulted in 5 to 10% lower naphthalene conversion, which could be due to the specifics of Ni-W-S hydrogenating phase formation on the zeolite surface [47]. It should be as well noted, that the decrease in the conversion of naphthalene with increasing temperature does not contradict the thermodynamic laws of the hydrogenation reaction of this compound.

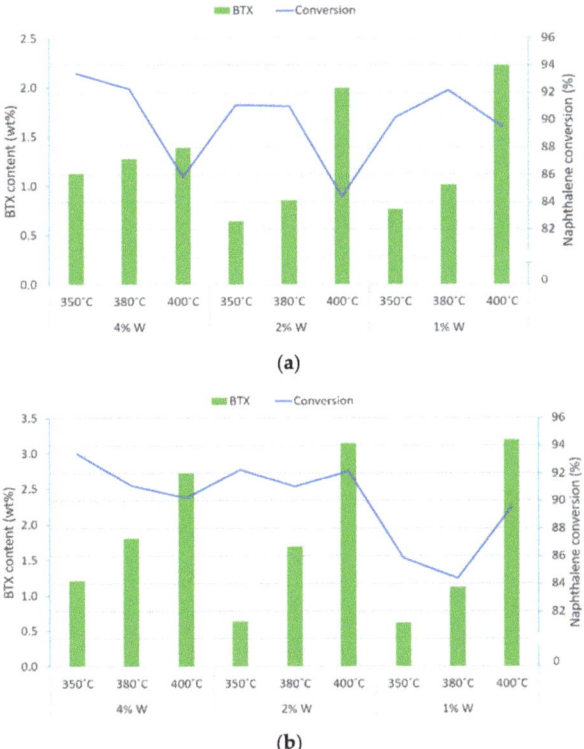

**Figure 1.** Naphthalene conversion versus tungsten molar ratio and reaction temperature, and benzene, toluene, and xylene (BTX) selectivity with (**a**) NiWS-HY(0.5%), (**b**) NiWS-HY(1%) catalysts at standard conditions P = 5 MPa, t = 5 h, and W:Ni ratio = 1:2.

It can be seen from the graph of Figure 1 that the highest conversion is achieved on the average at 380 °C and a tungsten content of 4 wt % [46]. No significant changes in naphthalene conversion versus NiWS-HY(0.5%) or NiWS-HY(1%) zeolite content were revealed.

The sum of the benzene, toluene, and xylenes recovered during the hydrocracking process is about 2 to 3.5%. The highest BTX yield were obtained at 400 °C and a weight content of 1% for zeolite and 4% for tungsten (Figure 2). Under these conditions, in a mixture of xylenes, the ratio m-xylene:p-xylene = 1:10, and o-xylene is formed in trace amounts. We used these results to identify the dependencies of BTX components of selectivity on temperature and tungsten content (Figure 2). As a result, the highest benzene content as compared to xylene and toluene is observed.

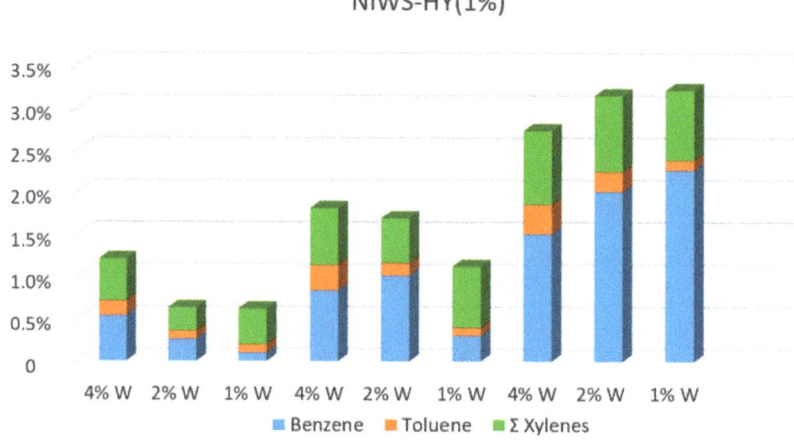

**Figure 2.** BTX (benzene, toluene, xylene) components selectivity in hydrocracking reaction of naphthalene solution with NiWS-HY(1%) catalysts at standard conditions P = 5 MPa, t = 5 h, and W:Ni ratio = 1:2.

At the same time, the main products of naphthalene hydrocracking under these conditions are tetralin and decalin. The higher their temperature, the lower their content in the reaction products, and the higher the BTX fraction content. The reaction is probably of consecutive behavior, with the tetralin acting as an intermediate. Scheme 1. If the hydrogenation rate is higher than the tetralin cracking rate, tetralin is accumulated and then hydrogenated to yield decalins. In a different scenario, tetralin is not generated at all [9,27].

Considering the hydrocracking results for a model mixture and preceding experience of the colleagues, the best hydrocracking results for the PFO fraction are expected when NiWS-HY(1%) is used since a high degree of feedstock conversion can probably be achieved when 1 wt % of zeolite is added [48], at 380 °C to 400 °C, and a tungsten content of 4% (Table 1).

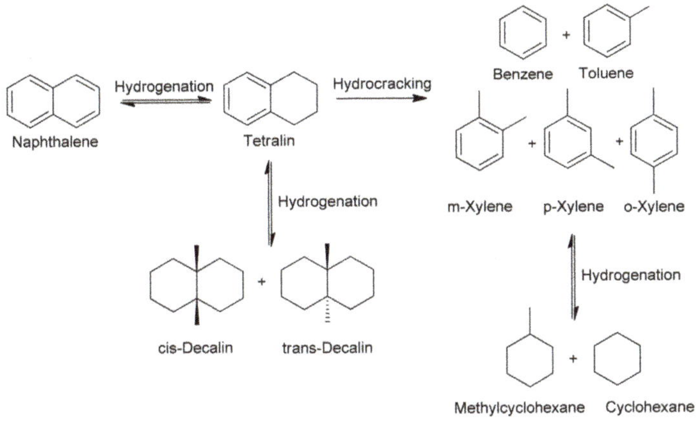

**Scheme 1.** Hydrocracking of pyrolysis fuel oil components into BTX.

**Table 1.** Detailed composition of naphthalene hydrocracking products for NiWS-HY(1%), W = 4% catalysts at standard conditions P = 5 MPa, t = 5 h, and W:Ni ratio = 1:2.

| Compound | Amount (wt %) | Compound | Amount (wt %) |
|---|---|---|---|
| Methylcyclopentane | 0.47 | Benzene | 2.84 |
| 3-Methylhexane | 0.39 | Heptane | 0.24 |
| Methylcyclohexane | 0.37 | Dimethylhexane | 0.17 |
| 4-Methylheptane | 0.5 | Toluene | 0.15 |
| 3-Methylheptane | 0.92 | Octane | 0.12 |
| m-Xynene | 0.72 | p-Xynen | 0.06 |
| o-Xynene | 0.04 | Decals | 3.4 |
| Butylbenzene | 0.2 | Methylindanes | 13.14 |
| Tetralin | 63.81 | - | - |

2.1.2. Pyrolysis Fuel Oil Hydrocracking with NiWS-HY Catalysts

The reactivity of in situ generated NiWS-HY(0.5%) and NiWS-HY(1%) catalysts during pyrolysis fuel oil hydroconversion was studied in an autoclave-type batch reactor at standard conditions p = 5 MPa, t = 5 h, and ratio W:Ni = 1:2, initial hydrogen pressure in the reaction system of 5.0 MPa, with intensive stirring. An HPR fraction with an upper boiling point of 330 °C, whose characteristics are given in *Materials and Methods* section, was used as a starting material.

To determine the optimal process temperature, we compared the active NiWS-HY(1%) and NiWS-HY (0.5%) catalysts with constant mass content W = 4%. Table 2 shows the BTX (benzene, toluene, xylene) selectivity between the reaction temperature and zeolite weight percentage. It was demonstrated that the in situ generated NiWS-HY(1%) catalyst is more active in the hydrocracking reaction than NiWS-HY (0.5%). The best selectivity of the BTX fraction was obtained at 380 °C. Maximum selectivity of BTX reaches 15 wt %.

When comparing the amount of the light fraction (with a boiling temperature up to 180 °C) generated at a different zeolite content, the predictably better results were obtained at a higher zeolite content (1 wt %) (Figure 3) [49]. The resulting light fraction also contains other valuable compounds.

**Table 2.** Dependence between component composition of the BTX fraction and temperature at standard conditions.

| Catalyst | Process Conditions | W (wt %) | BTX Composition (wt %) | | | |
|---|---|---|---|---|---|---|
| | | | Benzene | Toluene | Xylenes | ΣBTX |
| NiWS-HY(0.5%) | 350 °C | 4.0 | 3.3 | 2.5 | 0.8 | 6.6 |
| | 380 °C | | 7.7 | 2.2 | 0.95 | 10.85 |
| | 400 °C | | 5 | 0.7 | 1.1 | 6.8 |
| NiWS-HY(1%) | 350 °C | | 5.4 | 2.2 | 1.5 | 9.1 |
| | 380 °C | | 9.5 | 3.7 | 1.8 | 15 |
| | 400 °C | | 3.3 | 4.1 | 2.3 | 9.7 |

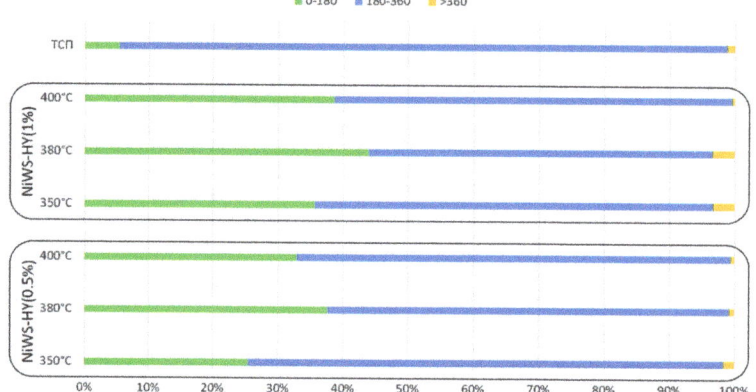

**Figure 3.** Fractional composition of HPR hydrocracking products with NiWS-HY(0.5%), NiWS-HY(1%) catalysts at constant mass content W = 4% and at standard conditions ($p$ = 5 MPa, t = 5 h, and W:Ni ratio = 1:2).

It is also noteworthy that PFO hydrocracking yields were 8–10 wt % in the reaction mass of cyclohexane as one of the benzene hydrogenation products. The cyclohexane is a non-toxic intermediate compound, which is widely used for the production of nylons and caprolactam. Currently, the greatest part of the world's cyclohexane production is based on benzene hydrogenation [50].

If we consider the dependence of the component composition of BTX reaction products (Table 3) on the cat/raw material ratio, we can note that the selectivity of benzene formation in the mixture increases with an increasing of wt % of tungsten, while selectivity of xylenes formation, on the contrary, increases with a decreasing of the amount of tungsten.

**Table 3.** Fractional composition of reaction products versus catalyst/feedstock ratio and zeolite content at 380 °C with NiWS-HY(0.5%), NiWS-HY(1%) catalysts at standard conditions P = 5 MPa, t = 5 h, and W:Ni ratio = 1:2.

| Catalyst | W (wt %) | Fractional Composition (wt %) | | | BTX Composition (wt %) | | | |
|---|---|---|---|---|---|---|---|---|
| | | 0 to 180 | 180 to 360 | >360 | Benzene | Toluene | Xylenes | ΣBTX |
| NiWS-HY(1%) | 4.0 | 44 | 52.7 | 3.3 | 9.5 | 3.7 | 1.8 | 15 |
| | 2.0 | 28.9 | 67.5 | 3.6 | 5.9 | 2.6 | 1.8 | 10.3 |
| | 1.0 | 24.2 | 72.4 | 3.34 | 3.9 | 2.5 | 2.1 | 8.5 |
| NiWS-HY(0.5%) | 4.0 | 37.7 | 61.5 | 0.8 | 7.7 | 2.2 | 1 | 10.9 |
| | 2.0 | 33 | 65.2 | 1.8 | 7.3 | 1 | 1.2 | 9.5 |
| | 1.0 | 31.8 | 68% | 0.2 | 5.4 | 1.8 | 2.1 | 9.3 |

The fractional composition of TSP after hydrocracking changes noticeably, the amount of the fraction with boiling points of 0–180 °C increases almost ten times, and it contains about 15 wt % of valuable components of BTX fraction (benzene-toluene-xylenes). The fraction with boiling points of 180–360 °C contains more than 30 wt % of decalins and tetralin.

## 2.2. Catalyst Properties

As is known, the activity of supported sulfide catalysts depends on the degree of sulfidation of the catalyst and the promoter/nickel ratio. The degree of sulfidation also depends on the temperature. The phases of nickel sulfide (NiS) and oxosulfide tungsten ($WO_xS_y$) are formed on a support structure at 300–400 °C. The higher the degree of interaction with the support, the more difficult the sulfidation of the oxysulfide phase and the formation of a mixed Ni-W-S phase are [49].

Based on the results of XPS, we revealed some dependences of the formed catalyst. We analyzed two catalysts obtained in situ in real raw material (pyrolysis fuel oil) at different zeolite/raw material ratios (0.5% and 1%) with catalyst sulfurization with elemental sulfur in the reaction medium (Tables 4 and 5). The best results for the selectivity of BTX fraction were showed with a tungsten content of 4 wt % at standard conditions ($p = 5$ MPa, $t = 5$ h, and W: Ni ratio = 1:2). The results of XPS obtained by us are consistent with the literature [51] since the signals of oxygen, carbon, sulfur, nickel, and tungsten correspond to characteristic signals of nickel-tungsten-sulfide catalysts. Promotion coefficient of NiWS-HY (0.5%) catalyst is 0.34, and promotion coefficient of NiWS-HY (1%) catalyst is only 0.19, Figure 4.

**Table 4.** XPS data for W4f, Ni2p, and S2p levels (catalyst obtained by decomposition of precursors with different concentrations of added zeolite) are 0.5% and 1%-HY in standard conditions.

| Element | | NiWS-HY(0.5%) | | | NiWS-HY(1%) | | | Status |
|---|---|---|---|---|---|---|---|---|
| | | Binding Energy, eV | Weight % | Atomic % | Binding Energy, eV | Weight % | Atomic % | |
| W4f | 4f 7/2<br>4f 5/2 | 33<br>35.1 | 65.6 | 3.5 | 33.2<br>35.3 | 52.53 | 2.77 | $WS_2$ |
| | 4f 7/2<br>4f 5/2 | 33.4<br>35.5 | 21.7 | 1.2 | 33.5<br>35.8 | 30.01 | 1.58 | $WO_xS_y$ |
| | 4f 7/2<br>4f 5/2 | 36.7<br>38.5 | 12.8 | 0.7 | 37<br>38.7 | 13.88 | 0.73 | $WO_3$ |
| Ni2p | 2p3/2<br>2p1/2 | 854.9<br>874.8 | 48.3 | 1.64 | 854.7<br>874.7 | 42.25 | 0.80 | NiS |
| | 2p3/2<br>2p1/2 | 857<br>877.6 | 34.5 | 1.17 | 856.6<br>877 | 28.09 | 0.53 | Ni-W-S |
| | 2p3/2<br>2p1/2 | 858.7<br>879.8 | 17.3 | 0.59 | 858.4<br>880 | 29.67 | 0.5% | NiO |
| S2p | 2p3/2 | 162.6 | 74.6 | 12.6 | 162.7 | 77.80 | 13.04 | $S^{2-}$ |
| | 2p3/2 | 163.9 | 14.2 | 2.4 | 164 | 12.94 | 2.17 | $S_2^{2-}$ |
| | 2p3/2 | 167.9 | 10.4 | 1.8 | 168 | 9.26 | 1.55 | $S^{6+}$ |

**Table 5.** Elemental composition of the catalyst surface according to XPS data.

| Catalyst | Atomic % | | | | | |
|---|---|---|---|---|---|---|
| | C | W | Ni | S | O | N |
| NiWS-HY(0.5%) | 63.146 | 5.311 | 3.390 | 16.922 | 11.232 | - |
| NiWS-HY(1%) | 66.637 | 5.273 | 1.891 | 16.763 | 9.436 | - |

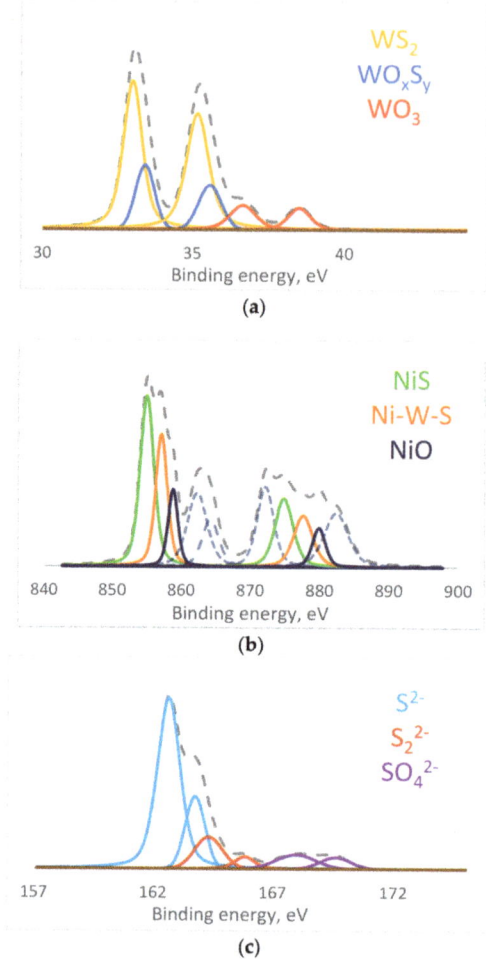

**Figure 4.** The (**a**) W 4f, (**b**) Ni 2p, and (**c**) S 2p XPS spectra of the NiWS-HY(0.5%) sample.

Based on the XPS results, it can be concluded that the active Ni-W-S phase is more easily formed on zeolite surface at mass content of 0.5%, which is reflected in the results of the catalytic experiment, and also in a larger amount of $WS_2$. It is also known to contribute to the hydrogenation process. But its activity is much lower than that of the phase Ni-W-S [52].

In an attempt to explain the experimental data on the activity of catalysts in hydrocracking, we studied the morphology of in situ formed catalysts isolated after the reaction using transmission electron microscopy (TEM). It should be noted that, according to TEM data, catalysts prepared in situ with the addition of zeolites directly to the reaction medium are spherical agglomerates (Figure 5). Moreover, in a sample with a lower zeolite content, the spherical shape is more pronounced, the average size of a spherical particle is from 0.18 to 0.2 µm, and in a catalyst sample with a high zeolite content, the average size of spherical particles is from 0.2 to 0.24 µm. The number of layers in sulfide packs (N) and the length of the stack (L) were determined manually. For the sample obtained with the addition of 0.5 wt % of zeolite, average N = 4.7, and L = 3.3, while obtained with the addition of 1 wt % of zeolite, N = 2.8, L = 6.2. We calculated the dispersion of particles (Ni-$WS_2$) according to the published method [51]. The dispersion of NiWS-HY (0.5%) catalyst 0.23 and NiWS-HY (1%) catalyst 0.2 was calculated.

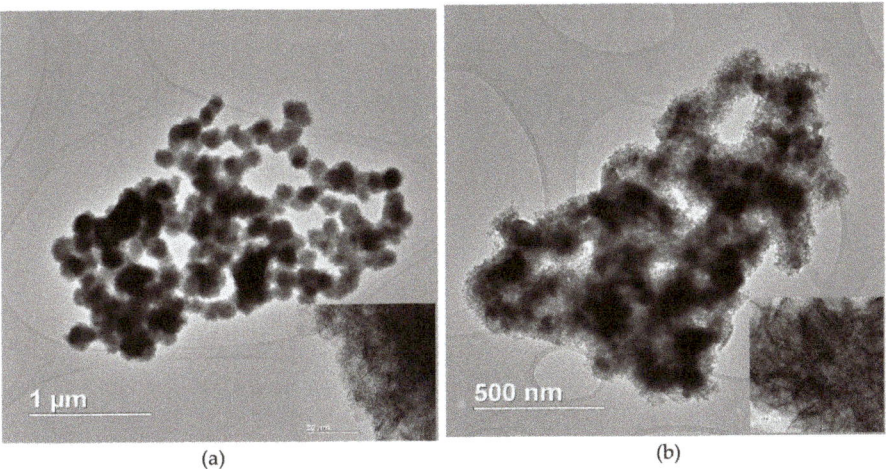

**Figure 5.** TEM micrograph of NiWS catalysts with (**a**) 0.5% and (**b**) 1% HY obtained in situ in standard conditions.

The size distribution of crystals of Ni-W, NiSx, etc., larger than 5 nm can be determined by analyzing the images of element maps. These element maps confirm the results of XPS and indicate that a larger amount of the hydrogenating Ni-W-S phase is formed under the condition of a smaller addition (0.5 wt %) of zeolite to the reaction medium. Thus, in this case, the results indicate the fact that the concentration of zeolite present in the mixture has a significant effect on the formation of the active phase.

Ni sulfide (NiSx) crystals are visible on TEM images and element maps of STEM-EDX and NiWS-HY (0.5%) (Figure 6 and NiWS-HY (1%) (Figure 7) hydrocracking catalysts. Particles range from about 5 to over 100 nm in diameter, all of them are surrounded with a Ni-W-S phase. As is known from the literature, a large amount of crystalline nickel sulfide does not contribute significantly to the hydrogenation process [53]. The distribution maps added to the article Appendix A (Figures A1 and A2).

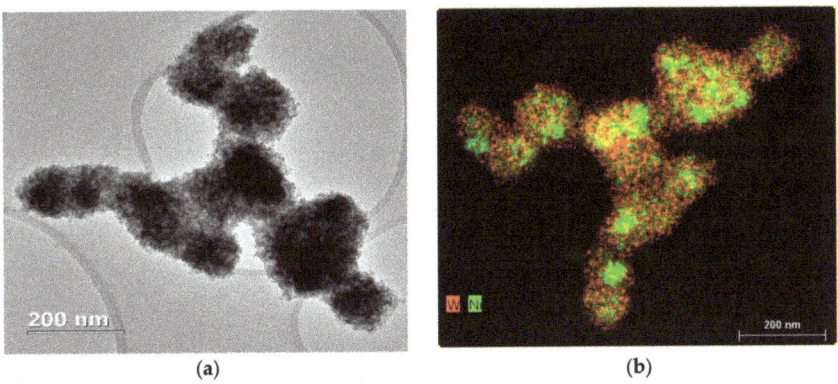

**Figure 6.** *Cont.*

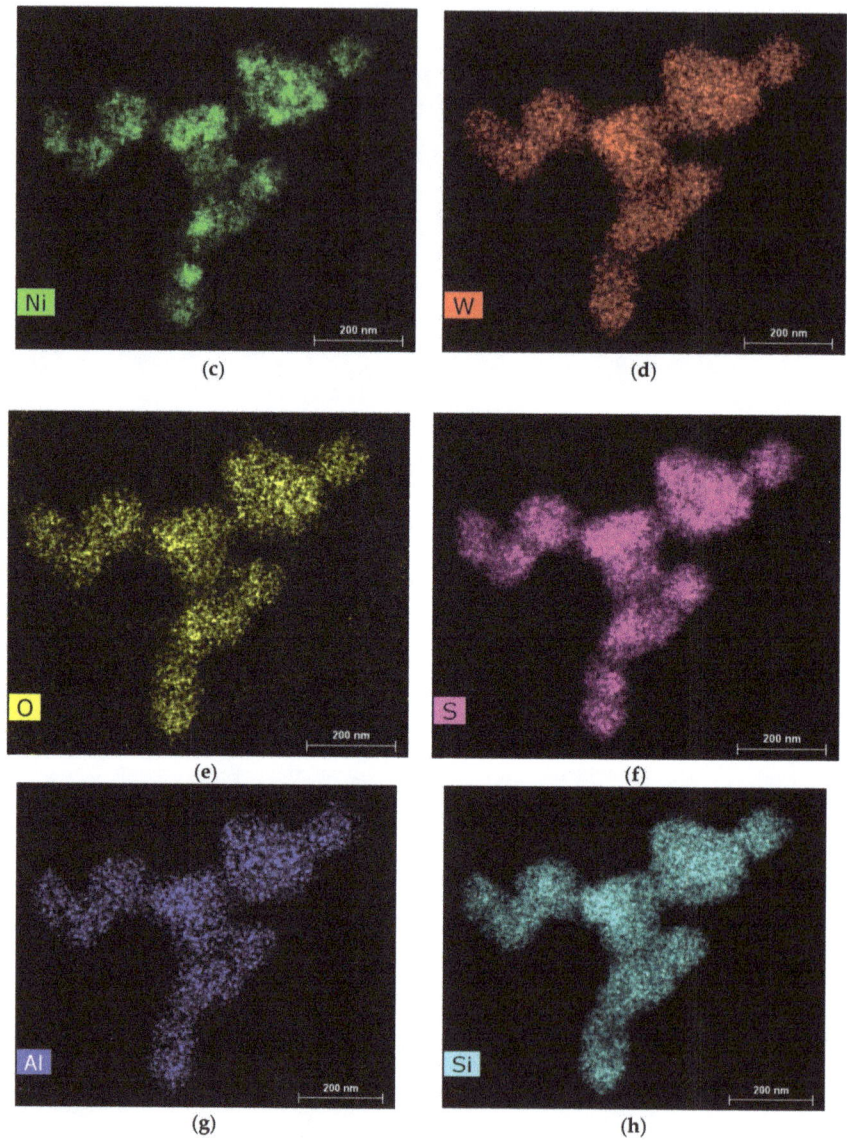

**Figure 6.** (**a**) TEM micrograph and distribution maps of NiWS-HY(0.5%) elements (**b**) Ni-W, (**c**) Ni, (**d**) W, (**e**) O, (**f**) S, (**g**) Al, (**h**) Si obtained in situ in standard conditions.

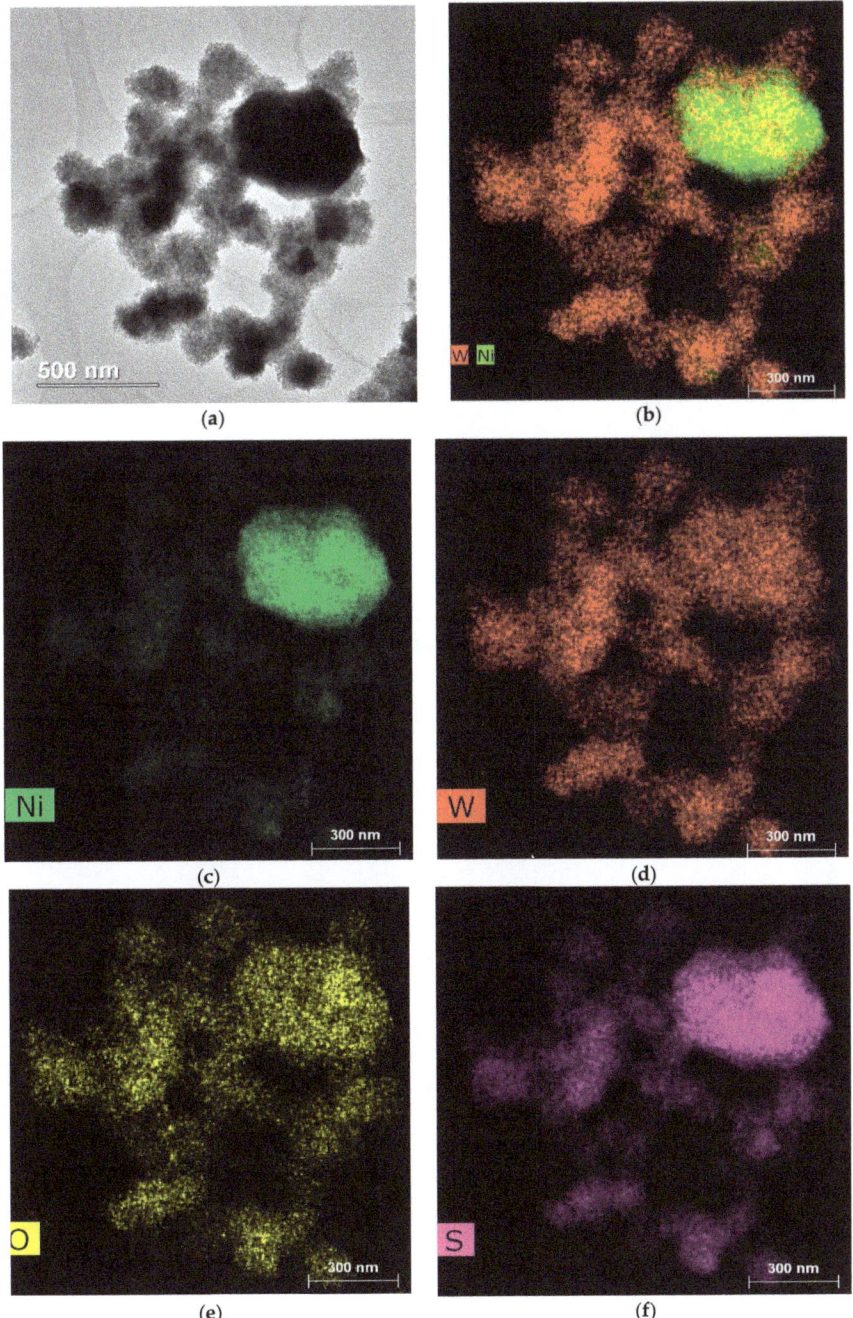

**Figure 7.** *Cont.*

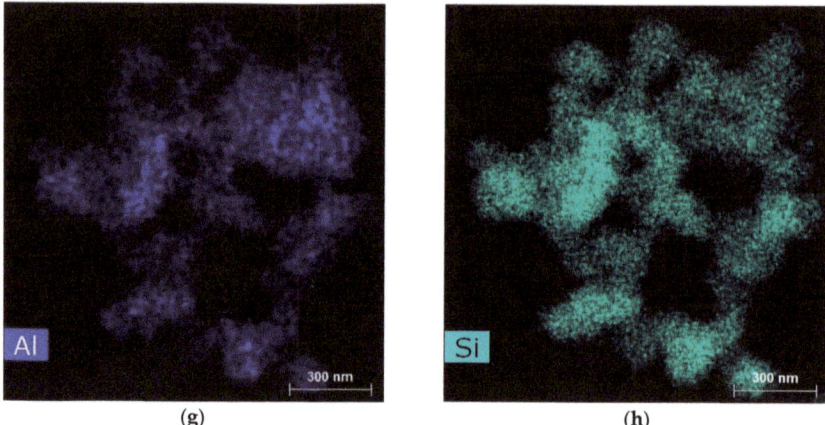

(g)                (h)

**Figure 7.** (a) TEM micrograph and distribution maps of NiWS-HY(1%) elements (b) Ni-W, (c) Ni, (d) W, (e) O, (f) S, (g) Al, (h) Si obtained in situ in standard conditions.

## 3. Materials and Methods

### 3.1. Materials

Reagents used are tungsten hexacarbonyl W(CO)$_6$ (99.99%, Aldrich), nickel(II 2-ethylhexanoate Ni(C$_7$H$_{15}$COO)$_2$ (78% solution in 2-ethylhexanoic acid, Aldrich). A solution of naphthalene (99%, Aldrich) in hexadecane (≥99%, Aldrich) was used as a substrate.

The fraction of the pyrolysis fuel oil was obtained from the pyrolysis fuel oil produced by Sibur-Kstovo (Kstovo, Russia) by vacuum distillation with the selection of IBP 170 °C fraction (residual distillation pressure-2 Torr). The output of the distillate fraction was 75.4 wt %, the boiling point reduced to atmospheric pressure was 330 °C. The composition shown in Figure 5 contains naphthalene (35.98%), its methyl-substituted analogs (10.95%), and biphenyl (27.97%) (Figure 8).

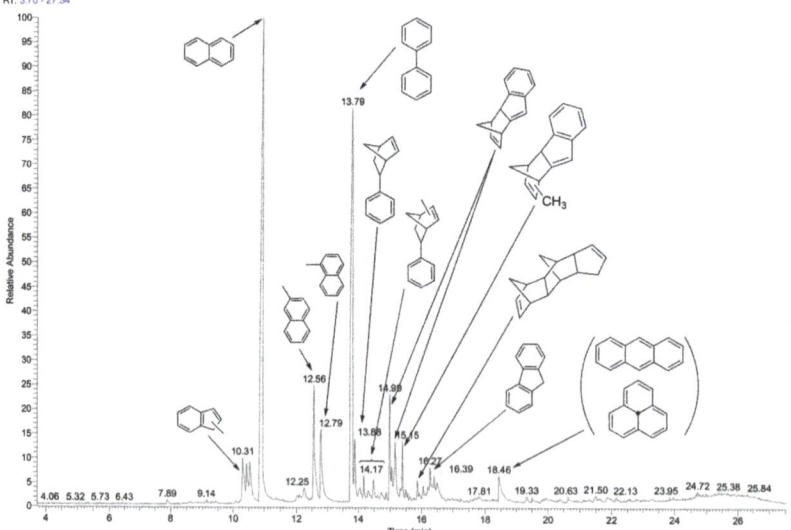

**Figure 8.** GC-MS plot of pyrolysis fuel oil fraction.

The mass chromatogram of the fraction was obtained on a Finnigan MAT 95 XL instrument (Varian VF-5 ms capillary column, length of 30 m, inner diameter of 0.25 mm, film thickness of 0.25 µm). All chemicals were used without preliminary purification.

*3.2. Catalyst Synthesis*

Sulfide catalysts of the NiWS-HY composition were prepared in situ in a hydrocarbon raw material. Low-soluble salts of hexacarbonyl tungsten $W(CO)_6$ and nickel(II)2-ethylhexanoate $Ni(C_7H_{15}COO)_2$ were used as precursors. To form the hydrogenating component of the catalyst in the form of tungsten sulfide promoted with nickel, 2.5 wt. % of elemental sulfur was added to the reaction mixture as a sulfiding agent. As the acidic component of the catalyst, a commercial wide-pore zeolite Y(CBV 600) with a three-dimensional porous structure and a channel diameter of 7.4 Å and cavities of 12 Å was chosen, which was added directly to the reaction medium. In zeolite molar ratio $SiO_2 / Al_2O_3 = 5.2$. According to TEM data particle size is 400–800 nm. Specific surface area found by the low-temperature adsorption of nitrogen is 530 m²/g. Total concentration of acid sites according to TPD of ammonia is 532 µmol/g [47]. The influence of the content of active metal and zeolite on the catalytic activity and selectivity of BTX production and the formation of the catalyst was also studied.

The structure and morphology of the catalysts obtained in situ were investigated by transmission electron microscopy (TEM) (JEOL, Tokyo, Japan), FEI Tecnai Osiris TEM at an accelerating voltage of 200 keV. For transmission electron microscopy (TEM), a FEI Tecnai Osiris TEM with an accelerating voltage of 200 keV was used.

To measure the (Ni)WS$_2$ dispersion, the average fraction of W atoms at the (Ni)WS$_2$ edge surface was calculated, assuming that the sulfide slabs were perfect hexagons [47,53]. The presence of a certain number of nickel atoms is not considered. Dispersion (D) was statistically evaluated by dividing the total number of W atoms at the edge surface (We), including corner sites (Wc), by the total number of W atoms ($W_T$) using the slab sizes measured in the TEM micrographs:

$$D = \frac{We + Wc}{W_T} \quad (1)$$

$$We = (6 \times n_i - 12) \times \overline{N} \quad (2)$$

$$Wc = 6 \times \overline{N} \quad (3)$$

$$W_T = \left(3 \times n_i^2 - 3 \times n_i + 1\right) \quad (4)$$

where $n_i$ is the number of W atoms along one side of the (Ni)WS$_2$ slab, as determined by its average length $\overline{L}$.

$$\overline{L} = \frac{\sum l_i}{n} \quad (5)$$

where $l_i$ is a length of each slab.

$$n_i = \frac{10 \times \frac{\overline{L}}{3.2} + 1}{2} \quad (6)$$

and

$$\overline{N} = \frac{\sum n_i N_i}{n}, \quad (7)$$

where $\overline{N}$ in the average stacking degree and $n_i$ is the number of stacks in $N_i$ layers.

The XPS analysis of catalysts was performed by ESCALAB MK2 electronic spectrometer (Vacuum Generators LTD) (Physical Electronics, Chanhassen, MN, USA). The samples were studied without heating and ion-beam treatment. The position of the lines of the elements was normalized to the position of C 1s carbon line of the hydrocarbon contamination of the surface. The survey spectrum was measured at an analyzer bandwidth of 50 eV and a scanning interval of 0.25 eV; the particular spectra of the elements were measured at a bandwidth of 20 eV and a scanning interval of 0.2 eV. Deconvolution of the spectra was conducted by non-linear least-square method using the Gaussian-Lorentzian function.

The absolute content of the Ni-W-S phase on the catalyst surface ($C_{NiWS}$) was calculated using the formula:

$$C_{NiWS} = \frac{[NiWS] * C_{Ni}}{100} \qquad (8)$$

where [NiWS] is the relative Ni content in the Ni-W-S phase, %; $C_{Ni}$ is the content of Ni atoms on the catalyst surface determined by the XPS method, at. %.

The absolute content of the $WS_2$ phase on the catalyst surface ($C_{WS2}$) was calculated from the formula:

$$C_{WS2} = \frac{[WS_2] * C_W}{100} \qquad (9)$$

where [$WS_2$] is the relative W content in the $WS_2$ phase, %; $C_W$ is the content of W atoms on the catalyst surface determined by the XPS method, at. %.

The promoter ratio in the active phase slab was determined:

$$(Ni/W)slab = \frac{C_{Ni-W-S}}{C_{WS_2}} \qquad (10)$$

*3.3. Catalytic Tests*

A number of experiments were carried out, first at a model mixture, to determine the best conditions for hydrocracking, and then at a fraction of pyrolysis fuel oil (PFO). A 10% solution of naphthalene in hexadecane was chosen as a model mixture. Hydrocracking was carried out in an autoclave-type batch reactor (Stainless steel high-pressure batch reactor). A sample of the catalyst (precursors W $(CO)_6$, $Ni(C_7H_{15}COO)_2$ and zeolite HY) was placed in it. The absence of the external diffusion limitations was checked in a separate test.

The sulfiding of the catalysts was carried out directly in the reaction medium by adding elemental sulfur in an amount of 2.5 wt. % in terms of raw materials. Then, the autoclave was sealed and filled with hydrogen to a pressure of 5.0 MPa, the reaction duration was 5 h. We varied the process temperature (350–400 °C), zeolite cont (0.5 and 1 wt %) and W (1–4 wt %) in the reaction medium, the ratio Ni/W = 2:1. The tungsten content in the raw material was calculated using the following formula:

$$W = \frac{m(W(CO)_6)M(W)}{M(W(CO)_6)m(сырья)} \times 10^6 \ [ppm] \qquad (11)$$

where $m(W(CO)_6)$ is the mass of tungsten hexacarbonyl dissolved in the hydrocarbon raw material, g; M (W) is molar mass of tungsten, 183.8 g/mol; $M(W(CO)_6)$ is molar mass of tungsten hexacarbonyl, 351.9 g/mol; m (raw material) is the mass of hydrocarbon raw material, g.

The start time of the experiment was considered the moment when the required reaction temperature was established. Upon completion of the reaction, the autoclave was cooled to room temperature and depressurized. The catalyst was separated from the products by centrifugation, and the liquid reaction products were analyzed by gas-liquid chromatography method.

*3.4. Products Characterization*

Qualitative analysis of liquid products was carried out using Agilent 6890 gas-liquid chromatograph with Finnigan MAT 95 XL mass spectrometric detector (Thermo Scientific) (Supelco, Bellefonte, PA, USA), Varian VF-5MS capillary column (30 m × 0.25 mm × 0.25 μm), with helium as carrier gas. The quantitative and fractional composition of liquid products (gasoline fraction < 180 °C, diesel fraction 180–360 °C, residual fraction > 360 °C) was determined by simulated distillation using a Kristallux 4000 M gas-liquid chromatograph with a flame ionization detector, capillary column SPB-1 (30 m × 0.25 mm × 0.25 μm) (Meta-Khrom, Yoshkar-Ola, Russia) with helium as carrier gas.

The naphthalene conversion (conversion, %) was calculated as the degree of conversion of the initial aromatic compound into hydrocracking products.

## 4. Conclusions

In this paper, hydrocracking of a pyrolysis fuel oil in a slurry reactor on in situ synthesized NiWS-HY catalysts from low-soluble precursors for the selective production of the BTX fraction was studied for the first time. The qualitative and quantitative composition of the hydrocracking products was studied. It was found that the yield of benzene-toluene-xylenes can reach up to 15% of the total mass of the products. An increase was also noted for the proportion of light products of hydrocracking.

Thus, we can conclude that NiWS-HY catalyst can be considered as a catalyst for the hydrocracking of pyrolysis fuel oil. Moreover, varying the process conditions led to a change in the selectivity for the products of BTX fraction.

The optimal conditions for the process of obtaining BTX fraction from pyrolysis fuel oil were determined. At 380 °C, with 4 wt % of tungsten, and 1 wt % and 0.5 wt % of zeolite, the amount of the product was much greater. The best result was obtained at 380 °C, with 4 wt % of tungsten, and 1 wt % of zeolite, where mass of BTX fraction was about 15%.

The catalyst obtained under these conditions was characterized by XPS method; it was found that with a lower content of zeolite more of the hydrogenating Ni-W-S phase was formed. This is also confirmed by the TEM results.

**Author Contributions:** Conceptualization, T.K., M.K., V.S., and A.M.; methodology, M.K.; investigation, T.K.; data curation, T.K.; writing—original draft preparation, T.K.; writing—review and editing, A.M. and V.S.; supervision, A.M.; project administration, A.M. All authors have read and agreed to the published version of the manuscript.

**Funding:** This research received no external funding.

**Acknowledgments:** This work was carried out within the State Program of TIPS RAS.

**Conflicts of Interest:** The authors declare no conflict of interest.

## Appendix A

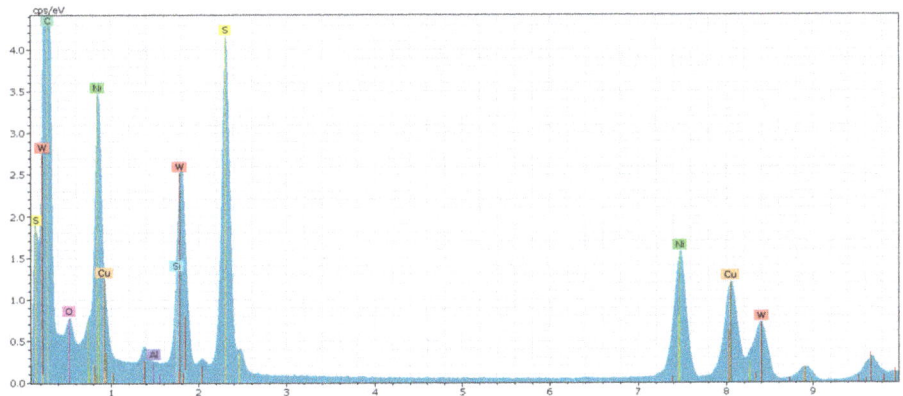

**Figure A1.** EDX spectrum of catalyst NiWS-HY (0.5%).

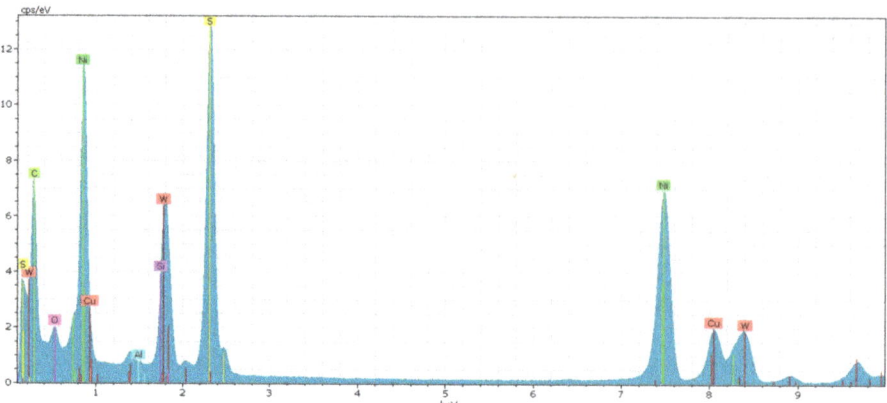

**Figure A2.** EDX spectrum of catalyst NiWS-HY (1%).

## References

1. Yan, J.D. Biomass to energy in china: Development status and strategic consideration. *Chem. Ind. For. Prod.* **2014**, *34*, 151–158.
2. Cho, M.H.; Jung, S.H.; Kim, J.S. Pyrolysis of mixed plastic wastes for the recovery of benzene, toluene, and xylene (BTX) aromatics in a fluidized bed and chlorine removal by applying various additives. *Energy Fuels* **2010**, *24*, 1389–1395. [CrossRef]
3. Xu, X.W.; Jiang, E.C. "BTX" from guaiacol HDO under atmospheric pressure: The effect of support and the carbon deposition. *Energy Fuels* **2017**, *31*, 2855–2864. [CrossRef]
4. Xu, N.; Pan, D.; Wu, Y.; Xu, S.; Gao, L.; Zhang, J.; Xiao, G. Preparation of nano-sized HZSM-5 zeolite with sodium alginate for glycerol aromatization. *React. Kinet. Mech. Catal.* **2019**, *127*, 449–467. [CrossRef]
5. Song, W.; Zhou, S.; Hu, S.; Lai, W.; Lian, Y.; Wang, J.; Jiang, X. Surface engineering of CoMoS nano-sulfide for hydrodeoxygenation of lignin-derived phenols to arenes. *ACS Catal.* **2019**, *9*, 259–268. [CrossRef]
6. Jung, S.H.; Cho, M.H.; Kang, B.S.; Kim, J.S. Pyrolysis of a fraction of waste polypropylene and polyethylene for the recovery of BTX aromatics using a fluidized bed reactor. *Fuel Process. Technol.* **2010**, *91*, 277–284. [CrossRef]
7. Xu, X.W.; Jiang, E.C.; Li, Z.Y.; Sun, Y. BTX from anisole by hydrodeoxygenation and transalkylation at ambient pressure with zeolite catalysts. *Fuel* **2018**, *221*, 440–446. [CrossRef]
8. Tamiyakul, S.; Ubolcharoen, W.; Tungasmita, D.N.; Jongpatiwut, S. Conversion of glycerol to aromatic hydrocarbons over Zn-promoted HZSM-5 catalysts. *Catal. Today* **2015**, *256*, 325–335. [CrossRef]
9. Laredo, G.C.; Pérez-Romo, P.; Escobar, J.; Garcia-Gutierrez, J.L.; Vega-Merino, P.M. Light cycle oil upgrading to benzene, toluene, and xylenes by hydrocracking: Studies using model mixtures. *Ind. Eng. Chem. Res.* **2017**, *56*, 10939–10948. [CrossRef]
10. Polliotto, V.; Livraghi, S.; Agnoli, S.; Granozzi, G.; Giamello, E. Reversible adsorption of oxygen as superoxide ion on cerium doped zirconium titanate. *Appl. Catal. A Gen.* **2019**, *580*, 140–148. [CrossRef]
11. Choi, Y.; Lee, J.; Shin, J.; Lee, S.; Kim, D.; Lee, J.K. Selective hydroconversion of naphthalenes into light alkyl-aromatic hydrocarbons. *Appl. Catal. A Gen.* **2015**, *492*, 140–150. [CrossRef]
12. Kim, Y.-S.; Cho, K.-S.; Lee, Y.-K. Structure and activity of $Ni_2P$/desilicated zeolite β catalysts for hydrocracking of pyrolysis fuel oil into benzene, toluene, and xylene. *Catalyst* **2020**, *10*, 47. [CrossRef]
13. Upare, D.P.; Park, S.; Kim, M.S.; Jeon, Y.-P.; Kim, J.; Lee, D.; Lee, C.W. Selective hydrocracking of pyrolysis fuel oil into benzene, toluene and xylene over CoMo/beta zeolite catalyst. *J. Ind. Eng. Chem.* **2017**, *46*, 356–363. [CrossRef]
14. Sato, K.; Iwata, Y.; Miki, Y.; Shimad, H. Hydrocracking of TEtralin over NiW/USY zeolite catalysts: For the improvement of heavy-oil upgrading catalysts. *J. Catal.* **1999**, *186*, 45–56. [CrossRef]

15. Ishihara, A.; Itoh, T.; Nasu, H.; Hashimoto, T.; Doi, T. Hydrocracking of 1-methylnaphthalene/decahydronaphthalene mixture catalyzed by zeolite-alumina composite supported NiMo catalysts. *Fuel Process. Technol.* **2013**, *116*, 222–227. [CrossRef]
16. Park, J.I.; Lee, J.K.; Miyawaki, J.; Kim, Y.K.; Yoon, S.H.; Mochida, I. Hydro-conversion of 1-methyl naphthalene into (alkyl)benzenes over alumina-coated USY zeolite-supported NiMoS catalysts. *Fuel* **2011**, *90*, 182–189. [CrossRef]
17. Ali, M.A.; Tatsumi, T.; Masuda, T. Development of heavy oil hydrocracking catalysts using amorphous silica-alumina and zeolites as catalyst supports. *Appl. Cataly. A Gen.* **2002**, *233*, 77–90. [CrossRef]
18. Silva-Rodrigo, R.; Calderón-Salas, C.; Melo-Banda, J.A.; Domínguez, J.M.; Vázquez-Rodríguez, A. Synthesis, characterization and comparison of catalytic properties of NiMo- and NiW/Ti-MCM-41 catalysts for HDS of thiophene and HVGO. *Catal. Today* **2004**, *98*, 123–129. [CrossRef]
19. Leyva, C.; Rana, M.S.; Trejo, F.; Ancheyta, J. On the use of acid-base-supported catalysts for hydroprocessing of heavy petroleum. *Ind. Eng. Chem. Res.* **2007**, *46*, 7448–7466. [CrossRef]
20. Leyva, C.; Rana, M.S.; Trejo, F.; Ancheyta, J. NiMo supported acidic catalysts for heavy oil hydroprocessing. *Catal. Today* **2009**, *141*, 168–175. [CrossRef]
21. Tiwari, R.; Rana, B.S.; Kumar, R.; Verma, D.; Kumar, R.; Joshi, R.K.; Garg, M.O.; Sinha, A.K. Hydrotreating and hydrocracking catalysts for processing of waste soya-oil and refinery-oil mixtures. *Catal. Commun.* **2011**, *12*, 559–562. [CrossRef]
22. Looi, P.Y.; Mohamed, A.R.; Tye, C.T. Hydrocracking of residual oil using molybdenum supported over mesoporous alumina as a catalyst. *Chem. Eng. J.* **2012**, *181*, 717–724. [CrossRef]
23. Nishijima, A.; Shimada, H.; Sato, T.; Yoshimur, Y.; Hiraishi, J. Support effects on hydrocarcking and hydrogenation activities of molybdenum catlysts used for upgrading coal-derived liquids. *Polyhedron* **1986**, *5*, 243–247. [CrossRef]
24. Shimada, H. Morphology, dispersion and catalytic functions of supported molybdenum sulfide catalysts for hydrotreating petroleum fractions. *J. Jpn. Pet. Inst.* **2016**, *59*, 46–58. [CrossRef]
25. Castillo-Villalón, P.; Ramírez, J.; Cuevas, R.; Vázquez, P.; Castañeda, R. Influence of the support on the catalytic performance of Mo, CoMo, and NiMo catalysts supported on $Al_2O_3$ and $TiO_2$ during the HDS of thiophene, dibenzothiophene, or 4,6-dimethyldibenzothiophene. *Catal. Today* **2016**, *259*, 140–149. [CrossRef]
26. Laredo, G.C.; Pérez-Romo, P.; Vega-Merino, P.M.; Arzate-Barbosa, E.; García-López, A.; Agueda-Rangel, R.; Martínez-Moreno, V.H. Effect of the catalytic system and operating conditions on BTX formation using tetralin as a model molecule. *Appl. Petrochem. Res.* **2019**, *9*, 185–198. [CrossRef]
27. Upare, D.P.; Park, S.; Kim, M.S.; Kim, J.; Lee, D.; Lee, J.; Lee, C.W. Cobalt promoted Mo/beta zeolite for selective hydrocracking of tetralin and pyrolysis fuel oil into monocyclic aromatic hydrocarbons. *J. Ind. Eng. Chem.* **2016**, *35*, 99–107. [CrossRef]
28. Yi, Y.; Jin, X.; Wang, L.; Zhang, Q.; Xiong, G.; Liang, C. Preparation of unsupported Ni-Mo-S catalysts for hydrodesulfurization of dibenzothiophene by thermal decomposition of tetramethylammonium thiomolybdates. *Catal. Today* **2011**, *175*, 460. [CrossRef]
29. Olivas, A.; Zepeda, T.A.; Villalpando, I.; Fuentes, S. Performance of unsupported Ni(Co,Fe)/$MoS_2$ catalysts in hydrotreating reactions. *Catal. Commun.* **2008**, *9*, 1317–1328. [CrossRef]
30. Beller, M.; Renken, A.; van Santen, R.A. *Catalysis: From Principles to Applications*; Wiley: Hoboken, NJ, USA, 2012.
31. Furimsky, E.; Massoth, F.E. Deactivation of hydroprocessing catalysts. *Catal. Today* **1999**, *52*, 381–495. [CrossRef]
32. Hagenbach, G.; Courty, P.; Delmon, B. Physicochemical investigations and catalytic activity measurements on crystallized molydbenum sulfide-cobalt sulfide mixed catalysts. *Catalyst* **1973**, *31*, 264–273. [CrossRef]
33. Topsøe, H.; Clausen, B.S. Importance of Co-Mo-S type structures in hydrodesulfurization. *Catal. Rev.* **1984**, *26*, 395–420. [CrossRef]
34. Hur, Y.G.; Kim, M.-S.; Lee, D.-W.; Kim, S.; Eom, H.-J.; Jeong, G.; Lee, K.-Y. Hydrocracking of vacuum residue into lighter fuel oils using nanosheet-structured $WS_2$ catalyst. *Fuel* **2014**, *137*, 237–244. [CrossRef]
35. Bellussi, G.; Rispoli, G.; Landoni, A.; Millini, R.; Molinari, D.; Montanari, E.; Moscotti, D.; Pollesel, P. Hydroconversion of heavy residues in slurry reactors: Developments and perspectives. *J. Catal.* **2013**, *308*, 189–200. [CrossRef]

36. Knyazeva, M.I.; Panyukova, D.I.; Kuchinskaya, T.S.; Kulikov, A.B.; Maximov, A.L. Effect of composition of cobalt-molybdenum-containing sulfonium thiosalts on the hydrogenation activity of nanosized catalysts in situ synthesized on their basis. *Pet. Chem.* **2019**, *59*, 1285–1292. [CrossRef]
37. Kennepohl, D.; Sanford, E. Conversion of athabasca bitumen with dispersed and supported mo-based catalysts as a function of dispersed catalyst concentration. *Energy Fuels* **1996**, *10*, 229–234. [CrossRef]
38. Alonso, G.; Chianelli, R.R. WS2 catalysts from tetraalkyl thiotungstate precursors and their concurrent in situ activation during HDS of DBT. *J. Catal.* **2004**, *221*, 657–661. [CrossRef]
39. Jeong, H.-R.; Lee, Y.K. Comparison of unsupported WS2 and MoS2 catalysts for slurry phase hydrocracking of vacuum residue. *Appl. Catal. A Gen.* **2019**, *572*, 90–96. [CrossRef]
40. Sizova, I.A.; Panyukova, D.I.; Maksimov, A.L. Hydrotreating of high-aromatic waste of coke and by-product processes in the presence of in situ synthesized sulfide nanocatalysts. *Pet. Chem.* **2017**, *57*, 1304–1309. [CrossRef]
41. Onishchenko, M.I.; Kulikov, A.B.; Maksimov, A.L. Application of zeolite Y-based Ni-W supported and in situ prepared catalysts in the process of vacuum gas oil hydrocracking. *Pet. Chem.* **2017**, *57*, 1287–1294. [CrossRef]
42. Hur, Y.G.; Lee, D.-W.; Lee, K.-Y. Hydrocracking of vacuum residue using NiWS(x) dispersed catalysts. *Fuel* **2016**, *185*, 794–803. [CrossRef]
43. Jeon, S.G.; Na, J.-G.; Ko, C.H.; Lee, K.B.; Rho, N.S.; Park, S.B. A new approach for preparation of oil-soluble bimetallic dispersed catalyst from layered ammonium nickel molybdate. *Mater. Sci. Eng. B.* **2011**, *176*, 606–610. [CrossRef]
44. Ben Tayeb, K.; Lamonier, C.; Lancelot, C.; Fournier, M.; Payen, E.; Bonduelle, A.; Bertoncini, F. Study of the active phase of NiW hydrocracking sulfided catalysts obtained from an innovative heteropolyanion based preparation. *Catal. Today* **2010**, *150*, 207–212. [CrossRef]
45. McDonald, W.; Friesen, G.D.; Rosenhein, L.D.; Newton, W.E. Syntheses and characterization of ammonium and tetraalkylammonium thiomolybdates and thiotungstates. *Inorg. Chim. Acta* **1983**, *72*, 205–210. [CrossRef]
46. Furimsky, E. *Catalyst for Upgrading Heavy Petroleum Feed*; Elsevier: Amsterdam, The Netherlands, 2007; p. 404.
47. Kniazeva, M.; Maximov, A. Effect of additives on the activity of Nickel-Tungsten sulfide hydroconversion catalysts prepared in situ from oil-soluble precursors. *Catalysts* **2018**, *8*, 644. [CrossRef]
48. Zhu, L.; Sun, H.; Fu, H.; Zheng, J.; Zhang, N.; Li, Y.; Chen, B.H. Effect of ruthenium nickelbimetallic composition on the catalytic performance for benzene hydrogenation tocyclohexane. *Appl. Catal. A Gen.* **2015**, *49*, 124–132. [CrossRef]
49. Hensen, E.J.M.; van der Meer, Y.; van Veen, J.A.R.; Niemantsverdriet, J.W. Insight into the formation of theactive phases in supported NiW hydrotreating catalysts. *Appl. Catal. A Gen.* **2007**, *322*, 16–32. [CrossRef]
50. Díaz de León, J.N.; Antunes-García, J.; Alonso-Nuñez, G.; Zepeda, T.A.; Galvan, D.H.; de los Reyes, J.A.; Fuentes, S. Support effects of NiW hydrodesulfurization catalysts from experiments and DFT calculations. *Appl. Catal. B Environ.* **2018**, *238*, 480–490. [CrossRef]
51. Eijsbouts, S.; Li, X.; Bergwerff, J.; Louwen, J.; Woning, L.; Loos, J. Nickel sulfide crystals in Ni-Mo and Ni-W catalysts: Eye-catching inactive feature or an active phase in its own right? *Catal. Today* **2017**, *292*, 38–50. [CrossRef]
52. Voorhoeve, R. The mechanism of the hydrogenation of cyclohexene and benzene on nickel-tungsten sulfide catalysts. *J. Catal.* **1971**, *23*, 243–252. [CrossRef]
53. Kasztelan, S.; Toulhoat, H.; Grimblot, J.; Bonnelle, J.P. A geometrical model of the active phase of hydrotreating catalysts. *Appl. Catal.* **1984**, *13*, 127–159. [CrossRef]

© 2020 by the authors. Licensee MDPI, Basel, Switzerland. This article is an open access article distributed under the terms and conditions of the Creative Commons Attribution (CC BY) license (http://creativecommons.org/licenses/by/4.0/).

MDPI
St. Alban-Anlage 66
4052 Basel
Switzerland
Tel. +41 61 683 77 34
Fax +41 61 302 89 18
www.mdpi.com

*Catalysts* Editorial Office
E-mail: catalysts@mdpi.com
www.mdpi.com/journal/catalysts

www.ingramcontent.com/pod-product-compliance
Lightning Source LLC
LaVergne TN
LVHW070544100526
838202LV00012B/369